Leibniz: Journal Articles on Natural Philosophy

LEIBNIZ FROM OXFORD

This series presents new translations of works by

G. W. Leibniz with introductions and notes,
for a broad philosophical readership.

Leibniz: Journal Articles on Natural Philosophy
Edited by Richard T. W. Arthur
Translations by Jeffrey McDonough, Lea Schroeder, Samuel Levey, Richard T. W. Arthur, Richard Francks (with Roger Woolhouse), and Tzuchien Tho

Leibniz: Dissertation on Combinatorial Art
Translated with introduction and commentary by
Massimo Mugnai, Han van Ruler, and Martin Wilson

Leibniz: Discourse on Metaphysics
Translated with introduction and commentary by
Gonzalo Rodriguez-Pereyra
Edited with an Introduction by Richard Arthur

Leibniz: Journal Articles on Natural Philosophy

Edited by

RICHARD T. W. ARTHUR

Translations by

JEFFREY K. McDONOUGH,
LEA AURELIA SCHROEDER, SAMUEL LEVEY,
RICHARD T. W. ARTHUR, RICHARD FRANCKS
(*with* ROGER WOOLHOUSE),
and
TZUCHIEN THO

OXFORD
UNIVERSITY PRESS

OXFORD
UNIVERSITY PRESS

Great Clarendon Street, Oxford, OX2 6DP,
United Kingdom

Oxford University Press is a department of the University of Oxford.
It furthers the University's objective of excellence in research, scholarship,
and education by publishing worldwide. Oxford is a registered trade mark of
Oxford University Press in the UK and in certain other countries

Published in the United States of America by Oxford University Press
198 Madison Avenue, New York, NY 10016, United States of America

British Library Cataloguing in Publication Data
Data available

Library of Congress Control Number: 2023931429

ISBN 978–0–19–284353–1

DOI: 10.1093/oso/9780192843531.001.0001

Printed and bound in the UK by
Clays Ltd, Elcograf S.p.A.

Contents

8. On Optical Lines and other matters: Excerpts from a letter to – –. G. W. Leibniz, *Acta eruditorum*, January 1689 81

9. A Sketch Concerning the Resistance of a Medium and the motion of heavy bodies projected in a resisting medium: G. W. Leibniz, *Acta eruditorum*, January 1689 85

10. An Essay on the Causes of the Celestial Motions: *Acta eruditorum*, February 1689 95

11. Observations on the Cause of Gravity and its Properties: Denis Papin, *Acta eruditorum*, April 1689 112

12. On the Isochronous Line along which a heavy body descends without acceleration, and on the controversy with the Abbé D. C.: G. W. Leibniz, *Acta eruditorum*, April 1689 118

13. On the Cause of Gravity, and defence against the Cartesians of the author's own view on the true laws of nature: G. W. Leibniz, *Acta eruditorum*, May 1690 123

14. A Reply to the articles that the illustrious J. B. published in the May issue of these *Acta*: G. W. Leibniz, *Acta eruditorum*, July 1690 135

15. An Opinion about the Motive Forces of Mechanics, offered by D. Papin against the objections of the most distinguished G. G. L.: Denis Papin, *Acta eruditorum*, January 1691 138

16. Addition to the Sketch on the Resistance of the Medium published in these *Acta* in February 1689: G. W. Leibniz, *Acta eruditorum*, April 1691 147

17. On the Line into which a flexible body curves itself under its own weight, and its remarkable usefulness for discovering any number of mean proportionals and logarithms: G. W. Leibniz, *Acta eruditorum*, June 1691 149

18. On Solutions to the Problem of the Catenary or Funicular, and to other problems proposed by the learned I. B. in the *Acta* of June 1691: G. W. Leibniz, *Acta eruditorum*, September 1691 156

19. On the Laws of Nature and true estimation of motive forces, against the Cartesians: a Reply to the arguments proposed by Mr P. last January in these *Acta*, p. 6: G. W. Leibniz, *Acta Eruditorum*, September 1691 162

20. General Rule for the Composition of Motions: G. W. Leibniz, *Journal des Sçavans*, September 1693 171

21. Two Problems Constructed by Mr Leibniz, employing the general rule of the composition of motions that he just published: G. W. Leibniz, *Journal des Sçavans*, September 1693 174

22. A Specimen of Dynamics for the disclosing of the admirable laws of nature concerning the forces of bodies and their mutual actions, and reducing them to their causes: G. W. Leibniz, *Acta eruditorum*, April 1695 176

23. A Short Note on p. 537ff. of the December *Acta* of 1695: G. W. Leibniz, *Acta eruditorum*, March 1696 191

24. An Excerpt from a Letter of G. G. L. that he wrote to a friend in favour of his physical hypothesis about the motions of the planets, once inserted in these *Acta* (Febr. 1689): *Acta eruditorum*, October 1706 195

25. Letters from Baron von Leibniz to Mr Hartsoeker, with the replies of Mr Hartsoeker: G. W. Leibniz and Nicolaas Hartsoeker, *Mémoires de Trévoux*, March 1712 201

26. Letter from Baron von Leibniz to Mr Hartsoeker, 12 July 1711: G. W. Leibniz, *Mémoires de Trévoux*, April 1712 214

Bibliography 217
Index of Names 225
Subject Index 228

Preface

In March 2015 I received an email from Paul Lodge informing me that he was preparing a volume of Leibniz's published philosophical journal articles for Oxford University Press, and asking if I would be interested in taking charge of a companion volume on his journal articles in natural philosophy, insofar as the two can be successfully distinguished. I accepted, and submitted a proposal to the acquisitions editor of OUP, Peter Momtchiloff; the present volume is the result. I invited various other scholars to help in the translations, and Sam Levey, Tzuchien Tho, and Jeffrey McDonough graciously volunteered their services in 2015, followed by Richard Francks in 2019, and Lea Aurelia Schroeder in 2020 after we received our contract from OUP (which was then reworked to include her). Richard Francks provided translations of **[4]**, **[5]**, **[6]**, and [7] (based on previous versions with the late Roger Woolhouse); Jeffrey McDonough, Lea Schroeder, and Samuel Levey collaboratively produced drafts for **[1]**, **[2]**, **[8]**, **[14]**, **[17]**, **[18]**, and **[23]**; Tzuchien Tho provided drafts for **[9]**, **[16]**, **[25]**, and **[26]**, and I provided the remainder, **[3]**, **[10]**, **[11]**, **[12]**, **[13]**, **[15]**, **[19]**, **[20]**, **[21]**, **[22]**, and **[24]**. All these drafts of the translations were complete by the Spring of 2021, and I then set about revising them and adding footnotes in consultation with the others, and writing the introduction.

A book of this kind cannot be achieved without help from outside, and it is a pleasure to thank Professor Stephan Meier-Öser of the University of Münster for his invaluable help in obtaining usable images of the Tables from the *Acta eruditorum* containing Leibniz's figures, and also to Osvaldo Ottaviani for his advice on this matter. For their helpful comments on the introduction I am grateful to Jeffrey, Tzuchien, and Richard, and also to Niccolò Guicciardini. I am also grateful to Paul Lodge and an anonymous reader for OUP for their helpful suggestions and corrections.

Richard Arthur

Toronto, 2022

Abbreviations

A	G.W. Leibniz. *Sämtliche Schriften und Briefe.* Ed. Akademie der Wissenschaften (Leibniz 1923–); cited by series, volume and page e.g. (A VI 2, 229), and occasionally by piece number, e.g. (A III 3, N. 269).
AT	*Oeuvres de Descartes,* 12 vols., Ed. Charles Adam & Paul Tannery. (Descartes 1964–76); cited by volume and page, e.g. (AT VIII 1, 71).
ESP	*Essais Philosophiques et scientifiques.* Ed. Lamarra et Palaia (Leibniz 2005).
GM	Gerhardt, *Leibnizens Mathematische Schriften* (Leibniz 1849–63); cited by volume and page, e.g. (GM II 157).
GP	Gerhardt, *Die Philosophischen Schriften von Gottfried Wilhelm Leibniz* (Leibniz 1875–90); cited by volume and page, e.g. (GP II 268).
L	Loemker, *Leibniz: Philosophical Papers and Letters,* (Leibniz 1976).
OCH	Huygens, *Œuvres complètes.* (Huygens 1888–1950).

Note on Texts and Translation

In all cases, the translations have been made from the original journal articles in the *Acta eruditorum*, the *Nouvelles de la République des Lettres*, the *Journal des Sçavans*, and the *Mémoires de Trévoux* (more properly, *Mémoires pour l'Histoire des Sciences & des Beaux-Arts*). Digital images of these articles may be found in the *Internet Archive* or in *Gallica*, and Gerhardt provided edited reprints in *Mathematische Schriften*, as noted. We have indicated the page breaks in the original articles in square brackets, thus [199–200]. Interpolations of missing words are made <thus>; explanatory interpolations and call-outs to figures are indicated [thus]. We have used 'ôr' to denote the Latin "or of equivalence" (translating sive or seu) to distinguish it from an 'or' denoting alternatives (in Latin, aut).

Names and titles: In the articles of the *Acta eruditorum* names of authors or antagonists in a debate were usually (although not always) abbreviated (especially in the article titles), thus "G. G. L." for "G. G. Leibniz" (i.e. Godefroi Guilliaume Leibnitz or Godefridus Guilelmus Leibnitius), or "Mons. L.", or "M. L.", for "Monsieur Leibniz", "M. D. C." for "Monsieur De Catelan", and so forth. This convention of using initials instead of names was used by editors to conceal identities in case there were issues of priority or conflict (sometimes leading to confusions on the part of readers, for example, on the identities of Leibniz and Tschirnhaus),[1] although the authors' identities are often revealed in the articles. We have retained the acronyms, the identities being clear from the footnotes, while translating "M." and "Monsieur" as "Mr.". We have not followed a second convention, however, that of italicizing all mentions of authors or sects, such as "*Huygens*" or "the *Cartesians*". Also, spellings of names were rather fluid in the seventeenth century. So we find Leibniz spelled variously as "Leibnits", "De Leibnis", and "de Leibnits", L'Hôpital as L'Hospital, and so forth; in all cases we have used the modern spellings.

[1] See Beeley (2020) for an account of how the Scottish mathematician John Craig managed to conflate Tschirnhaus (under the acronym "D.T.") with Leibniz, with the result that when Leibniz corrected Tschirnhaus in print, Craig accused the conflated author of contradicting himself! (On Craig, see the pertinent footnote in text [22].) Walther Ehrenfried von Tschirnhaus (1651–1708) was a German mathematician and friend of Leibniz's in Paris, who, as we shall see below, published some (defective) parts of Leibniz's method of quadrature without the latter's permission.

Seventeenth-century conventions of politeness have been retained where these are not too awkward: thus the honorific *Cl.*, short for *Clarissimus*, has been retained and expanded as "the most distinguished", whereas *Dn.* ("the learned") is closer to "Mr." in English, and we have rendered it accordingly.

Accents: In seventeenth century French, accents are not always given where they should be in modern French. Thus in some of the original language titles we reproduce one sees *replique, mecanique, ecrit* and *Fevrier* instead of *réplique, mécanique, écrit*, and *février*; but "*où il prétend soûtenir*...".

Technical terms: In this period there was considerable variation in the use of terms such as "velocity", which had not yet acquired their precise modern meanings. Consequently we have for the most part given literal translations of the various technical terms, rendering, for example, *velocitas* as "velocity" and *celeritas* (in French, *vitesse*) as "speed", even though they are used interchangeably. There was not yet at this point a precise concept of mass; we have translated its precursor *moles* (in French, *grandeur*) as "magnitude". Descartes's *quantitas motus* ("quantity of motion") was by this time a technical term, equated with the product of the magnitude of a body with its speed. Also in use was Galileo's concept of *gradus celeritatis* ("degree of speed"): if, for instance, a body were to persist in motion for a time with the degree of speed it had at the end of an acceleration, it would then have a uniform velocity in a straight line with that degree of speed. Other terms such as *impetus* ("impetus") and *vis* ("force") were still used with diverging interpretations. Leibniz was the originator of the distinction of "transcendental" from "geometrical" quantities (in his paper *De geometria recondita* of 1686), although he actually called them *transcendent* rather than *transcendental*; we have followed modern usage.

Further discussion of seventeenth-century applied mathematics can be found in the following note.

Some Notes on Seventeenth-century Mathematics

It may be useful to say a few words on seventeenth-century mathematics. On the classical model, based on the theory of proportions laid out in Euclid's *Elements*, quantities are portrayed comparatively, thus obviating the need for constants. Thus one would say that in a uniform motion "the spaces are as the times", rather than that the distance over the time is constant, or that $s = ct$, where c is a constant. Similarly, in motion under a central force, instead of saying that a planet is attracted according to the inverse square of its distance, one says that the attractions on planets are "in reciprocal duplicate proportion to their distances"; we have translated this as "inversely as the squares of their distances". Likewise, the periodic times of the planetary orbits according to Kepler's Third Law are said to be "in sesquialterate ratio to the mean distances", meaning $(T_1 : T_2)^2 :: (R_1 : R_2)^3$, or, in algebraic terms, $T = kR^{3/2}$. One can see instances of this geometric paradigm in the representation of the product of two quantities by a rectangle with sides of lengths proportional to the quantities concerned. Similarly, there was a requirement of dimensional homogeneity, according to which all the terms appearing in an equation should be of the same dimension, so that they would all have geometric counterparts. Thus in an equation constants would be supplied in order to preserve homogeneity: $ax^2 + b^2x = c^3$, etc. It was customary, moreover, to write the squares (though not higher powers) as products, so this would be written $axx + bbx = c^3$. In this period, however, the geometric paradigm was losing its pre-eminence under the advance of algebraic geometry, and one finds ratios such as $bb{:}q$ treated as simple quotients bb/q. Notation, moreover, was not yet standardized, so that what we would write as $(x^2 - 2y^2)/r^3$ Leibniz would sometimes write as $xx - 2yy,: r^3$, where the comma indicates that what precedes it is treated as one expression, and the colon represents division.

Moreover, the introduction of constants in algebraic formulas will seem unsystematic from a modern standpoint: typically, they are introduced as is found convenient for the solving of equations, often serving a double purpose, and the necessity for constants of integration, for example, was only gradually appreciated. An example of the double purpose use is provided by Leibniz's constant a in the *Tentamen*, which is both the constant of proportionality between elementary triangles and moments of time in Kepler's Area Law ($r^2d\varphi=adt$), and also the latus rectum of the ellipse ($a=b^2/q$). Two other features of seventeenth-century mathematics that will seem slightly strange to the modern eye: in geometrical diagrams the abscissa is drawn vertically and the ordinate line horizontally, contrary to the modern convention; and in Leibniz's case, if not universally, where we would write M_1, M_2 etc. he writes ${}_1M$, ${}_2M$—even though on the accompanying figures, confusingly, the same points are often rendered with suffixes and not prefixes: M_1 for ${}_1M$ and so on.

Finally, in the momentaneous representation used by Newton and Leibniz, one considers elements of the various quantities x produced in a moment of time dt (Leibniz) or o (Newton): these are called "elementary" quantities dx (Leibniz), or "moments" of quantities (o times the fluxion of x for Newton). Thus Leibniz represents velocity by the ratio or quotient $dx{:}dt$, or simply as dx (tacitly, in a given moment dt), where dx is an element of space. Newton, however, allowed variation of quantities within a moment, whereas Leibniz insisted that there was no acceleration in the elements, and this is crucial for understanding the difference between his *solicitations* and Newtonian accelerations. A solicitation ddx (with the element of time assumed constant) integrates directly to velocity, $v = \int dv = \int ddx$, where on a modern understanding $v = \int dv = \int (dv/dt)\, dt = \int (d^2x/dt^2)\, dt$.[2] Also, Leibniz's differential equations are phrased entirely in terms of differentials, or differences of quantities: there is no concept of a derivative in this period, nor is there yet the modern concept of a function.

Leibniz explains some of this in the *Tentamen*. In §5 of that essay he presents an all-too-brief sketch of the status of the infinitesimals he uses there, a paragraph he subsequently refers to as his "Lemmas on Incomparables". The *Tentamen* is therefore a pioneering work, in which Leibniz provides one of the first examples of a differential equation, $brdr=\theta a\surd[ee-pp]$, as well as the first published instance of a differo-differential equation, $bdrdr + brddr = 2pa\theta dr{:}\surd[ee-pp]$. The equation he derives from these,

[2] See Bertoloni Meli (1993, 75–91), Bos (1974–75), and Guicciardini (1999, 152–63) for a fuller discussion of the intricacies of seventeenth-century mathematics.

$ddr = bbaa\theta\theta - 2aaqr\theta\theta,:bbr^3$, given all the equivalences noted above (and substituting h for his a as the Kepler constant, with a denoting only the latus rectum), would come out in modern notation as $d^2r/dt^2 = h^2(1/r^3 - 2/ar^2)$, which, unappreciated as it was by his contemporaries, is mathematically correct for an inverse square force.[3]

[3] See Aiton (1972b, 125–51) for a thorough analysis of Leibniz's differential equation.

Introduction

'Anyone who knows me only through my published works does not really know me', wrote Leibniz to his friend Vincent Placcius[1]—and given the vast quantity of unpublished manuscripts comprising his *Nachlass*, we can certainly appreciate the truth of that remark. Indeed, much recent Leibniz scholarship has focussed on this unpublished material, plumbing it for insights and anticipations of modern ideas—so much so that at this point we are in danger of swinging to the other extreme, and losing sight of the immediate and very significant impact Leibniz had on the philosophy and science of his own time. For although he did not bring many of his book projects to completion, Leibniz nevertheless published well over a hundred articles on a wide array of topics in the emerging scientific journals of his time. The present volume, presenting in English translation a selection of the articles Leibniz published on what was then termed natural philosophy, including topics in optics, mechanics, and cosmology, together with its sister volume comprising Leibniz's published articles of a more general philosophical nature, is intended to redress that balance.[2]

Of course, that distinction between 'natural' and 'more general' philosophy is very artificial, as is also that between Leibniz's scientific papers and those which would be classified as mathematical. Nevertheless, the texts selected for his volume hang together quite naturally, and make for a coherent volume. We have tried to select those articles that are nearer the heart of Leibniz's contributions to natural philosophy, omitting, for instance, early articles on time-keeping in portable clocks, on a curiosity of natural history, and on the discovery of phosphorus.[3] Many of these necessarily involve the application of Leibniz's new mathematical methods to the problems concerned, but we have

[1] Leibniz to Vincent Placcius, 21 February 1696 (Leibniz 1768, (ed. Dutens), VI, 1, p. 65).

[2] The present volume is a companion volume to a second set of translated articles under the editorship of Paul Lodge, which offers a selection of published articles of a more general philosophical nature.

[3] The paper on time-keeping in portable clocks was written while Leibniz was still in Paris, '*Extrait d'une Lettre de M. Leibniz*...' (Leibniz 1675), and appeared in English translation in the *Philosophical Transactions* the following month. The paper on 'the figure of a deer with extraordinary antlers' came out in the *Journal des Sçavans* of 5 July (Leibniz 1677a), and the article on the discovery and properties of phosphorus by Crafft and Brandt in the same journal the following month (Leibniz 1677b).

Leibniz: Journal Articles on Natural Philosophy. Richard T. W. Arthur, Oxford University Press.
DOI: 10.1093/oso/9780192843531.003.0001

not attempted to chart the whole development of his calculus.[4] The main topics covered are as follows: Leibniz's work in optics, on the fracture strength of materials, and on motion in a resisting medium, and his pioneering applications of his calculus to these issues by construing them as mini-max and inverse tangent problems; his critique of the Cartesian measure of motive force, 'quantity of motion', and his proposal of a different way of estimating force to replace it, issuing in his measure mv^2; his proposed theory of celestial motions and gravitation, and derivation of the inverse square law; challenge problems concerning the isochronous curve and the catenary; a sample of his work on gaming theory; and his critique of atomism.

Leibniz's provocative articles on certain of these matters engendered controversies: the Abbé Catelan tried to defend Descartes and the Cartesians against Leibniz's charges; Denis Papin also responded, and criticized Leibniz's theory of gravitation on behalf of the Cartesians; and Hartsoeker sought to defend himself and others from Leibniz's critique of atomism.[5] The relevant published papers of all three controversies are translated here, including those of Catelan ([**4**] and [**6**]), Papin ([**11**] and [**15**]), and Hartsoeker ([**25**]), allowing the reader to gain a stronger impression of the intellectual context in which Leibniz's innovations were received.

The first piece in the collection is Leibniz's publication on optics in the then newly established journal, the *Acta eruditorum*.[6] He had been approached in December 1681 by Christoph Pfautz, one of the journal's editors, who had requested that Leibniz submit a lead article for the first issue in 1682. For this Leibniz sent the *De vera proportione circuli*,[7] a ground-breaking treatment of infinite series expansions for various curves, including the alternating infinite series expansion for the ratio of the area of a circle to that of a square on its

[4] This separation between Leibniz's more nearly mathematical papers and those on natural philosophy is of course the most artificial aspect of our selection, since he developed his calculus through applying it to problems in optics and physics. Even so, the present selection has only five articles in common with a volume of twenty-six of Leibniz's mathematical articles published in French translation by Marc Parmentier, *La naissance du calcul différentiel* (Leibniz 1995), namely, [**8**], [**12**], [**14**], [**17**], and [**18**] in this volume, corresponding to VI, VII, VIII, X, and XI in his.

[5] Biographical details for these three thinkers will be given below. Catelan (as discussed in footnote 31 below) was by no means a philosopher of first rank, but his known association with Malebranche required Leibniz to respond in the hopes of influencing Malebranche's views. Denis Papin, similarly, was associated with Huygens, but was a respected natural philosopher in his own right, and Leibniz had a long correspondence with him. Although Hartsoeker had his own idiosyncratic views, his importance for Leibniz was as an advocate of atomism, a view shared by such notables as Huygens and Newton.

[6] By this time he had already published six papers in the *Journal des Sçavans*, In addition to the three papers mentioned in footnote 3 above, Leibniz had also published two notes on mathematical topics, and another on a 'fuming liquid' ('*Extrait de deux lettres écrites à l'auteur du Journal,...*' Leibniz 1681), which he had previously discussed with Günther Christoph Schelhammer (on whom, see footnote 15).

[7] *De vera proportione circuli ad quadratum circumscriptum in numeris rationalibus expressa*, [On the True Proportion of the Circle to the Circumscribed Square Expressed in Rational Numbers], *Acta eruditorum*, February 1682, pp. 41–6; French translation in Leibniz 1995, 61–81.

diameter (i.e. $\pi/4$), and the first appearances in print of the mathematical term 'transcendent' and of an equation with an unknown in the exponent, x^x (i.e. x to the power of x).[8] The optics paper ([1] in our collection) appeared the following month. In it Leibniz gave a neat derivation of the laws of reflection and refraction of light by appeal to the principle of minimal path, something that he had discussed with Christiaan Huygens in their correspondence.[9] It also provided an opportunity for him to present a striking practical application of his 'new method of maxima and minima'—although, having whetted the appetite of his readers, he did not reveal how the calculation was done using his differential calculus, saving that for his famous *Nova Methodus* paper of 1684.[10] As he later explained to Pfautz, he did not claim any priority for his use of the principle of minimal path itself, which he ascribed to Fermat and Snell, and before them ancient authors on catoptrics,[11] but saw himself as 'using this principle of metaphysics (of the shortest path) to bring about an agreement with the mechanical principle of the composition of motions' (A III 3, 806 ff.). It should be noted, however, that while Fermat interpreted his minimal principle as giving the path of least time, Leibniz interprets it instead in terms of the path offering least difficulty, the '*via facillima*', where a greater resistance of the medium results in a greater velocity.[12] Fermat, and following him Pardies and Ango,[13]

[8] In fact, due to a delay in the delivery of the manuscript, it only appeared in the second, February, issue. See O'Hara (forthcoming) for discussion.

[9] See Huygens to Leibniz, 11 January 1680 (A III 3, N. 4), and Leibniz to Huygens, 5 February 1680 (A III 3, N. 22). For an informative discussion of the 'principle of optimality' that forms the backbone of this paper, see McDonough (2022).

[10] *Nova methodus pro maximis et minimis, itemque tangentibus, quae nec fractas, nec irrationales quantitates moratur, et singulare pro illis calculi genus* [A New Method for Maxima et Minima, as well as tangents, unhindered by either fractions or irrational quantities, and a singular kind of calculus for finding them], *Acta eruditorum*, October 1684, pp. 467–73. This paper was composed in haste to secure his priority in response to his friend Ehrenfried Walther von Tschirnhaus's publications (discussed further below in this Introduction).

[11] 'Rather, what took Fermat many pages, I unify by a compendium of the two linelets by an application of my calculus (for Snell's demonstration is no longer extant), but I show that the ancients already used the same technique in Catoprics' (letter to Pfautz, 14 May 1683; A III 3, 806 ff.). See O'Hara (forthcoming, ch. 1).

[12] In so doing, Leibniz was agreeing with Descartes's physical reasoning, albeit with the qualification that for Descartes light travelled instantaneously, and only its determination followed the same laws as would a real velocity. Newton (in his *Principia* of 1687) would also hold that light moves faster in a denser medium. Huygens, however, provided a physical basis for Fermat's opposing assumption that light would travel faster in the rarer medium with his wave theory of light (Huygens 1690). See Sabra (1981, ch. 8) for discussion.

[13] Ignace-Gaston Pardies (1636–1673) was a French Catholic priest and natural philosopher, author of *Dissertatio de Motu et Natura Cometarum* [Dissertation on the Motion and Nature of Comets] (Bordeaux, 1665), *Discours du mouvement local* (Paris, 1670), and *La Statique* (Paris, 1673). In his manuscript *Traité complet d'Optique*, unpublished at the time of his death, he advocated a wave theory of light. This work was studied by Pierre Ango (1640–1694), a fellow priest and teacher at the College of La Flèche, who published parts of Pardies' work in his *L'Optique divisée en trois livres* [Optics divided into three books] in 1682. Ango differed from Pardies, however, in considering the speed of light to be greater in rarer media. See Sabra (1981, 195–7) for an illuminating account.

had supposed on the contrary that light would move more slowly in a denser medium.[14]

During this period Leibniz had kept up a vigorous correspondence on scientific matters with two of his acquaintances from his time in Paris, his compatriot Günther Christoph Schelhammer,[15] and the French natural philosopher Edme Mariotte, an avid experimentalist. This is the context for the second paper in our collection. Schelhammer wrote to Leibniz in 1681 that he was composing a book on hearing, and asked his friend for his observations on sound. Leibniz complied, sending Schelhammer detailed explanations of his views on acoustics in his letters of February–March 1681 and January 1682 (A III 3, N. 182 and N. 311). This contribution was duly acknowledged by Schelhammer in his book on hearing, *De auditu* (1684), in the section on the propagation of sound. According to Leibniz this propagation took place by means of the sympathetic vibrations of the tiny particles of the air, assumed as elastic, so that sound was conveyed through the air to the ear at a constant speed and at a constant pitch (corresponding to the sympathetic resonances of the air particles), independent of the volume of the source.[16]

Here the assumption of the elasticity of the air was crucial to the derivation of this novel result, and an assumption he knew would be well received by Mariotte, since the insistence on the inherent elasticity of all matter was a signal point of agreement between them.[17] So when Mariotte brought up the topic in their correspondence, Leibniz sent him in a letter of August 1681

[14] Pierre Fermat (1607–1665) developed his views on optics in critical reaction to Descartes's *Diotprique* in his correspondence with Claude Clerselier in the late 1650s.

[15] Schelhammer (1649–1716) was part of Leibniz's circle of friends in Paris, and is likely the inspiration for the character 'Gallutius' in Leibniz's dialogue *Pacidius Philalethi*, written as he left London for The Hague in 1676 (see Leibniz 2001, 129). Schelhammer returned to Germany in 1677, where he received a doctorate in medicine, became professor of botany in Helmstedt and married Maria Sophia Conring, daughter of Hermann Conring, the distinguished professor of medicine at Helmstedt. (Schelhammer later became professor of anatomy, surgery, and botany in Jena in 1689, and then full professor of medicine at Kiel in 1695). He and Leibniz remained friends for life, and it was Schelhammer's controversy with Johann Christoph Sturm that prompted Leibniz to intervene on his side and write and publish his *De ipsa natura* in 1698.

[16] See James O'Hara (forthcoming, ch. 1) for a full account of this exchange of letters. As O'Hara reports, Leibniz composed his thoughts on acoustics into a little treatise with the title '*De soni generatione...; excerpta ex Epistolis G. G. L.* [On the generation of sound...: excerpts from a letter of G. G. L.]' (LH XXXVII 1, Bl. 18–19), planned for publication in the *Acta*, but which never did appear.

[17] Edme Mariotte (*c.* 1620–1684) was a French Abbé and an indefatigable deviser of experiments in natural philosophy, who had been elected to the *Académie des Sciences* in 1668. After moving to Paris in 1670, he wrote four essays on physics, and the second of these, *De la nature de l'air* [On the Nature of Air] (1676), contains his statement of the Boyle–Mariotte law, according to which the volume of a gas varies inversely as its pressure. He wrote on a great variety of topics in optics, collision theory, heat absorption, and ballistics, made significant contributions to aerodynamics (noting that the resistance of air varies as the square of velocity), and discovered the blind spot in the visual field caused by the position of the optic nerve.

(A III 3, N. 269) a summary of what he had written on acoustics for Schelhammer. The same idea of the elasticity of matter lay behind the two authors' agreement on the fracture strength of materials. In fact, Mariotte had apprised Leibniz of his theory that materials are comprised of elastic fibres or filaments which bend before breaking, in a letter of 28 April 1678 (A III 2, N. 157). Accordingly, in July 1682 he sent Leibniz his critique of Galileo's theory in the *Discorsi* about the breaking of beams.[18] By assuming an absolutely rigid beam and a sudden break, Galileo had arrived at a proportionality factor of ½ for the fracture strength, whereas based on the assumption of the elasticity of the beam, Mariotte had calculated it as ¼. Leibniz recalculated it as $^1/_3$, and after an exchange of letters, Mariotte, in his letter of 5 June 1683, accepted Leibniz's calculation,[19] a result he was afterwards able to confirm experimentally. This is the context for, and indeed the content of, Leibniz's second paper in this collection, *Demonstrationes novae de resistentia solidorum* [New Demonstrations about the Resistance of Solids] published in the *Acta eruditorum* of July 1684.[20] In 1686 Mariotte also published the result in his *Traité du Mouvement des Eaux* [Treatise on the Motions of Water], and historians of science now refer to it as the 'Mariotte–Leibniz' theory of the strength of materials.[21] It was also in 1684 that Leibniz published in the *Acta eruditorum* his rules for the differential calculus in his ground-breaking *Nova Methodus*, to be followed two years later by his *De geometria recondita* [On a Recondite Geometry], explaining their application to quadrature.[22]

[18] Galileo Galilei, *Discorsi e Dimostrazioni Matematiche intorno à due nuove scienze* [Discourses and Mathematical Demonstrations relating to Two New Sciences], First Day (1638, 49–65). See the translation of Stillman Drake (Galilei 1974).

[19] Mariotte wrote: 'I will make an experiment as soon as possible to see if it is necessary to take a third of the thickness instead of a half' (A III 3, 832; quoted from O'Hara (forthcoming, introduction)).

[20] We should also mention a paper on statics Leibniz published in the *Acta* of 1685 which we have chosen not to include here (Leibniz 1685), whose title translates as: 'A Geometrical Demonstration of a Rule Accepted in Statics that was recently called into question, concerning the moments of heavy bodies on inclined planes, and an elegant solution of the case proposed in the *Acta* of November 1684 (p. 512), concerning a globe lying on two planes making a right angle together, determining how much it presses on each of the planes' (GM VI 112–16).

[21] See O'Hara's commentary (forthcoming, introduction) for references to the theory as the 'Mariotte-Leibniz theory' by Todhunter and Truesdell. See McDonough (2022) for a revealing analysis of Leibniz's argument and the idealizing assumptions on which his calculation is based.

[22] The *Nova methodus de maximis et minimis* appeared in the *Acta eruditorum* of October 1684, and *De geometria recondita* in the *Acta eruditorum* of June 1686. Notoriously, Leibniz's exposition in the *Nova Methodus* was very obscure to his contemporaries, not helped by typographical errors and his reticence to describe his differentials as infinitely small. As noted above, Leibniz was provoked into publishing his rules for the calculus by Tschirnhaus' publishing of what Leibniz considered to be insufficiently understood parts of his method of quadrature in his *Methodus datae figurae, rectis lineis & Curva Geometrica terminatae, aut Quadraturam, aut impossibilitatem ejusdem Quadraturae determinandi* [A Method for determining either the Quadrature of a given figure, bounded by straight lines and a geometric curve, or the impossibility of the same quadrature] in the *Acta eruditorum* of October 1683, which Leibniz had sought to correct in his article *De dimensionibus figurarum inveniendis*

Realizing the need to assert his priority in other results he had achieved, Leibniz finally decided to publish his views on the correct measure of force that he had discovered in 1678.[23] His seminal paper *Brevis Demonstratio Erroris memorabilis Cartesii & aliorum circa legem naturæ*... [A Brief Demonstration of a Notable Error by Descartes and others concerning a Law of Nature...] appeared in the *Acta eruditorum* of March 1686. This was his first salvo in what was to become a long controversy with the Cartesians. Leibniz must have realized that the paper would create a stir, since (as he acknowledged) the criticisms he levelled at Descartes and his followers applied equally to other contemporary leading lights, such as Fabri,[24] Dechales,[25] and Borelli.[26] But he probably never imagined that debate about the true measure of force would still be raging into the second half of the eighteenth century. In one respect Descartes had already been shown to be in error, and the public identification of this error had marked Leibniz's entry point into physics in the early 1670s.[27] This was the simultaneous publication by Huygens, Wallis, and Wren, and later Mariotte, of papers showing that Descartes had been mistaken in his claim that the quantity of motion—the product of a body's mass and speed (scalar velocity)—is always conserved, so that the sum of the products of these quantities for two bodies would be the same before and after a collision. In particular, in the collision of two inelastic bodies some quantity of motion

[On finding the measures of figures] of the *Acta eruditorum* of May 1684, with an *Additio* in December of the same year addressing the efforts of Johann Christoph Sturm (GM V 123–6, 126–7; see Parmentier, (Leibniz 1995), 82–92, 93–5, for French translations).

[23] For a full account of this episode, see Michel Fichant's *La réforme de la dynamique* (Leibniz 1994). Fichant includes the full text of the *De corporum concursu* with a thorough commentary, as well as transcriptions of related manuscripts. See also François Duchesneau's *La dynamique de Leibniz* (1994), and Garber and Tho (2018).

[24] Honoré Fabri (or Fabry) (1608–1688) was a Jesuit mathematician and natural philosopher who held the position of theologian in the Vatican basilica for 30 years. He wrote works in philosophy, mathematics, physics, astronomy, and biology, and his contributions were highly respected by his peers, in particular Mersenne and Leibniz.

[25] Claude François Milliet Dechales (or De Chales) was a Jesuit teacher and missionary, who taught in Paris, Lyon, Chambéri, Marseilles, and Turin. His *Cursus seu mundus mathematicus* (Leiden: Anissoniana, 1674) was a much-read work incorporating practical applications of mathematics in mechanics, optics, geography, architecture, music, navigation, and hydrostatics.

[26] Giovanni Alfonso Borelli (1608–1679) was an Italian natural philosopher, trained in mathematics, who made important contributions to mechanics and especially, under the stimulus of Marcello Malpighi, to its application in the science of life (biomechanics). His main work was the *De Motu Animalium*, published posthumously in 1680–81, but at this time Leibniz would have known his works on the motions of Jupiter's moons (1666), his *De vi percussionis* (Bologna, 1667), and *De motionibus naturalibus a gravitate pendentibus* (1670).

[27] For a detailed account of Leibniz's early physics, see Beeley (2018). Manuscripts on the laws of impact were read to the Royal Society by Wallis in November 1668, Wren in December 1668, and Wallis in January 1669, and published in the *Philosophical Transactions* in 1669. Huygens' conclusions (drawn from his *De motu corporum ex percussione*, which was only published posthumously) were also published in the March 1669 issue of the *Journal des sçavans* (OCH XVI, 179–81). Mariotte gave his account in his *Traité de la percussion* (Mariotte 1673).

would always be lost, even if the sum of the quantity of motion *in a particular direction* is still conserved, as those authors all agreed. Nevertheless it was widely agreed that the force of a body's motion is proportional to the body's quantity of motion,[28] and it was this proposition that was the focus of Leibniz's critique.

Leibniz's positive argument is not that experiment shows that the correct measure of motive force is mv^2 (the measure he had discovered in 1678, of what he would call *vis viva*), even if this measure of force is implicit in his acceptance of Galileo's result that the speed acquired by a body in falling through a height of 4 units is twice that acquired by the same body falling through 1, that is, that h is proportional to v^2. That would have been to make this measure of force an accident of this particular set of circumstances.[29] Indeed, in the case of bodies on either side of a balance, Leibniz acknowledges that their force is equal to their quantity of motion, but declares this to be an accident of that particular case. His argument rests rather on his claim to have identified the correct way of estimating force generally, namely 'by the quantity of the effect that it can produce'. In identifying this force or 'energy' as equivalent to the ability to do work, correctly identifying its dimensions and positing its universal conservation, Leibniz has good claim to have discovered the First Law of Thermodynamics that historians of science have customarily credited to later thinkers. But his insight that a body can possess such a force as a power to bring about such an effect, even while not actually moving, went right over the heads of his contemporaries. For them force was always understood as an impulsive force associated with a body's motion, and therefore as a *present* tendency to change place in a given direction from where the body is now.[30]

[28] A measure of how widespread this consensus was can be seen in Samuel Clarke's riposte to Leibniz in the Fifth Reply in their controversy, sent on 29 October 1716, but which apparently did not reach Leibniz before his death on 14 November. He accuses Leibniz of 'great confusion and inconsistency' in upholding the conservation of active force while at the same time acknowledging that the quantity of motion is not always conserved. For, he claims concerning the former, 'this impetus, or relative impulsive active force of bodies in motion, is evidently both in reason and experience, always proportional to the quantity of motion' (fn. to §93–95; Alexander 1956, 121). Newton evidently agreed with Clarke on this score, holding that God would have to replenish active force in the world from time to time in the face of its loss in inelastic collisions; see his *Opticks* (Newton 1704, Query 31); in (Alexander 1956, 180).

[29] François Duchesneau makes a similar point (Duchesneau 1994, 140). We should also note that Huygens had in fact identified the quantity mv^2 as being conserved in the collisions of perfectly hard bodies. But he saw this as specific to that particular circumstance, and not as applying to all the bodies of our regular experience. See Huygens (1669).

[30] As Dennis Des Chene has explained 'For Descartes, motion is always entirely actual, the instantaneous rupture of a body from its neighbors. It is, moreover, *from* a definite place, but never by nature to a place: it has . . . a direction but no *terminus*' (Des Chene 1996, 256). In contrast, Leibniz

This understanding of motion can be seen to be at the root of the reaction by François Abbé de Catelan on behalf of the Cartesians, published in the *Acta eruditorum* in September 1686 ([4] in our collection). As Leibniz appears to have realized, Catelan was not a scholar of the first rank, but his closeness to Malebranche required a deferential response.[31] For Catelan, and for many others, it was axiomatic that a swifter motion is one in which a given distance is covered in less time, or a greater distance in the same time.[32] In such a conception there is no recognition that in a non-uniform motion, such as that of a falling body, this swiftness—the speed with which the whole motion is accomplished—must be distinguished from the speed acquired as a result of the fall—what we would call the body's final velocity. Holding onto the older conception of speed proved a stumbling block for Descartes; indeed, it had also been so initially for Galileo, who circumvented it by a geometric argument that in a uniform acceleration the distance covered would be equal to that covered by the same body moving uniformly with half the 'degree of speed' it would have acquired by the end of the fall (Galileo's Mean Speed Theorem).[33] So when Catelan writes that the quantities of motion are proportional to bodies' masses and speeds, and that in equal times bodies 'always cover distances proportional to those speeds', he does not recognize that this would only apply to bodies in uniform motion, and that these 'speeds' cannot be equated with the (final) velocities acquired by bodies as a result of falling under gravity.[34] Perhaps as a result of this confusion, he tries to restrict the Cartesian conservation principle to motions impressed in equal times, so that the same body

argues in this article that 'force should be calculated . . . by future effect. However, it seems that force or power is something real in the present, and future effect is not', entailing, accordingly, that there is more to body than the Cartesians allowed.

[31] All that we now know of the Abbé de Catelan has been assembled and commented on in an article by André Robinet (1958). Catelan was close to, and perhaps secretary for, Malebranche until the latter's polemic with Régis in 1694, after which 'he completely disappeared from the republic of letters' (1958, 293). Robinet considers Catelan's scientific qualification to be 'a huge misunderstanding', and remarks of his various interventions: '(T)he shadows of narrow obedience to Descartes, the clouds of outmoded or misunderstood formulations, the veils of obstinacy and bad faith, only serve to throw into greater relief the replies of his adversaries' (1958, 289).

[32] See, for instance, Catelan's second response to Leibniz below: 'if one [body] has taken half an hour to move, and the other a quarter of an hour, we will certainly conclude that the motion of the second is double that of the first'.

[33] Thus 'degree of speed' became an established term of art, and we see Leibniz availing himself of it. Since it was agreed that there could be no motion in an instant, strictly speaking there could be no such thing as an instantaneous velocity. It was Newton who first freely talked of velocity at an instant, although he preferred the term 'fluxion'.

[34] Indeed, as has been argued by Damerow et al. (1992; see esp. ch. 3), Galileo himself was limited by his own inconsistent reliance on Aristotelian speed. For further discussion of these difficulties concerning 'speed' in a non-uniform motion, see also Jullien and Charrak (2002, esp. 37–8) and Arthur (2016).

covering a greater distance in the same time might be said to have the greater force of motion.

Leibniz has little difficulty responding to this objection in his reply of February 1687 ([**5**] in our selection), pointing out that the Cartesians interpreted the conservation of quantity of motion, equated with force, to hold quite generally, and that consequently it was perfectly sufficient to have shown that it fails in a particular case. Also, as Galileo had shown by his analyses of motion down inclined planes, the time of descent could be varied as one wished, and the speed acquired would still be proportional to the height through which the heavy body had fallen, a result he had confirmed through repeated experiments. So the time of fall is quite irrelevant, as is also the way in which gravity acts to produce this effect: all that matters is the speed acquired by falling through a given height. In his first reply to Catelan, Leibniz extends his analysis to Descartes's rules of collision, with the stated intent of persuading Nicolas Malebranche[35] of their falsity. The consideration of the conservation of forces in collisions, in fact, is how Leibniz had first persuaded himself of the necessity for introducing his new measure of force in the *De corporum concorsu*, which also depended on a consideration of continuity through all the various cases of collision. Thus in his reply to Catelan he also mentions how certain of Malebranche's rules, like the Cartesian rules he was correcting, violate continuity considerations—a criticism that he will amplify in the extract from his famous letter to Malebranche published as a paper in July of 1687, '*Extrait d'une Lettre de M. L. sur un Principe Général...*'.[36]

In his comment on Leibniz's reply of June 1687 ([**6**] in our collection), Catelan misreads Leibniz as deriving a contradiction from his Cartesian and Galilean premises—Descartes's principle 'that it takes as much force to raise a body of one pound to a height of 4 feet as to raise one of four pounds to a height of 1 foot', Galileo's law of fall of heavy bodies, and the formula for the

[35] Leibniz knew Nicolas Malebranche (1638–1715) from his time in Paris, and maintained a correspondence with him thereafter. In his two-volume work *De la recherche de la vérité*, Malebranche attempted to synthesize Descartes's thought with that of St Augustine, initially upholding the Cartesian account of motion. But in his *Des lois de la communication des mouvements* (*Laws of the Communication of Motions*) of 1692 he corrected some of Descartes's rules of collision, while still endorsing his law of the conservation of the quantity of motion (Malebranche, *Œuvres complètes*, Tome XVII-1, 50). Finally, as a result of Leibniz's interventions, he admitted in a 1700 re-edition of the *Lois* that the conservation law does not hold (*Des lois de la communication des mouvements*, [1700], *Œuvres complètes*, Tome XVII-1, 73).

[36] '*Extrait d'une Lettre de M. L. sur un Principe Général, utile à l'explication des loix de la nature, par la considération de la Sagesse Divine; pour server de réplique à la réponse du R. P.M.* [Extract from a letter of Mr. [Leibniz] on a General Principle useful for explaining the laws of nature through consideration of the Divine Wisdom; to serve as a reply to the response of the Reverend Father Malebranche]', *Nouvelles de la République des lettres*, Juillet 1687, 744–53 (GP III, 51–5).

measure of the quantity of motion. But, as Leibniz notes in his response to it in September 1687 ([7] in our collection), 'far from finding any contradiction in them, I used them myself to show the error of the Cartesian principle'—that is, the principle of the universal conservation of quantity of motion. In fact, Catelan's painstaking explanation of things that 'are so clear that in the end you have to accept them' only serves to lay bare the poverty of his understanding of the issues involved. He fails to see that his understanding of motions as needing to be compared in equal times is a consequence of his (Aristotelian) conception of speed as the overall swiftness with which a certain distance is covered. As we have seen, this is inadequate for the analysis of non-uniform motion such as falling under gravity, where a distinction has to be made between the degree of speed attained by a body at the end of its fall and a simple quotient of the distance by the time.

To drive home this deficiency in Catelan's understanding of the non-uniform motion of a body falling under gravity, and of equal increases of motion in equal times, Leibniz submits a challenge problem for the Cartesians: '*To find the line of descent along which a heavy body descends uniformly and approaches the horizon by equal amounts in equal times.*' This is the problem of finding the isochronous line—the curve traced by a heavy body falling under gravity in such a way that its velocity vertically downwards remains constant. It was the first of several challenge problems that Leibniz and his correspondents would submit over the years to demonstrate the superiority of their mathematical methods in physics. A successful solution was submitted by Christiaan Huygens in September 1687 (and published the following month (Huygens 1687)), but Leibniz did not publish his own solution, the *De linea isochrona* ([12] in our collection), until April 1689.

That publication followed a hiatus in Leibniz's publishing activity caused by his absence on his long journey through Germany, Austria, and into Italy between October 1687 and March 1690. The purpose of the trip was, in his capacity as court counsellor (*Hofrat*), to research the history of the Welfs (or Guelphs) for his patron, the Duke of Hanover, Ernst August.[37] His intensive examination of the archival records in Munich and Augsburg delivered the vital discovery of a dynastical connection between the Duke's (Hanoverian)

[37] Ernst August (1629–1698) was Elector of Hanover and father of the future George I of Great Britain. He became Elector on the death in 1679 of his brother Johann Friedrich (1625–1679), who had invited Leibniz to the Hanoverian court as a junior Hofrat and unofficial scientific adviser. Ernst August was much less sympathetic to Leibniz's intellectual pursuits than his brother had been, however, insisting that he complete the history of the house of Braunschweig-Lüneburg. See Antognazza (2009) for details of his employment, and a very readable account of his journey through the German states to Italy and back (2009, 281–319).

house of Braunschweig-Lüneburg and the Italian house of Este. But to prove it he would need to consult the archives of the Este family in Modena. Obtaining this permission was a protracted affair, and Leibniz had written to ask Francesco de Floramonti, the Hanoverian representative in Venice, to establish the necessary contact with the Este court in Modena. Impatient with the slowness of this process, Leibniz set out from Munich before receiving permission from the Duke to extend his trip, arriving in Vienna in May 1688. There he remained for nine months, all the while awaiting a reply from Floramonti. Permission to access the Modena archives was not finally granted until the very end of 1689, after he had already been in Italy for eight months,

Characteristically, Leibniz put all this time to good use. During his sojourn in Vienna, besides working hard on church reunification with Bishop Rojas and attending to various political matters on behalf of the Duchy, he wrote several notable philosophical tracts, including the (unfinished) *Specimen Inventorum*, a summary of his metaphysics that appears to have been drafted for publication.[38] Leibniz had been able to catch up on the latest news from the scientific world with the help of the Emperor's librarian, Daniel von Nessel, who even brought him books and manuscripts from the library to his lodgings when he was indisposed with influenza for several weeks at the end of 1688, including the issues of the *Acta eruditorum* that had come out since his departure in the Fall of 1687. This was how he had come across a review (by Christoph Pfautz) in the June issue of the *Acta* (pp. 304–15) of Isaac Newton's masterwork, the *Principia*, whose first edition had been published in 1687.[39] Newton's treatment there of various issues Leibniz had himself worked on—the causes of celestial motions, catoptric or dioptric lines, and the resistance of the medium—made him realize that he needed to get some of his ideas into print before they were pre-empted by other scholars. So he immediately set about quickly writing three papers of his own on these topics,[40] sending them off for publication in the *Acta eruditorum*, where they appeared in quick succession in January and February 1689.

The first of these was *De lineis opticis et alia* [On Optical Lines and other matters] ([**8**] in our collection), in which Leibniz describes how these 'optical

[38] A VI 4, 1615–30, 315, 320; reproduced with English translation in (Leibniz 2001, 302–33).

[39] Newton (1687). Pfautz gives a fairly accurate overview of the main topics and contentions of the *Principia* but without any mathematics.

[40] In fact, Leibniz had written a fourth paper at the time that he intended to publish in the *Acta*, *Tentamen de Legibus Naturae Mundi* [Essay on the Laws of Nature in the World] (LH 35, 10, 4 f 1.v-1), but which he presumably decided not to submit. See Bertoloni Meli (1993, 123), who describes it as being '20 numbered paragraphs on dead and living force, composition of motion, the law of continuity, cohesion, the impact laws, elasticity, the non-existence of the vacuum and of atoms.'

lines' were discovered first by Huygens, then by Newton; and how he had come to discover them by a different route, as he explains in the paper. The second [**9**], *G. G. L. Schediasma de Resistentia Medii et Motu projectorum gravium in medio resistente* [A Sketch concerning the Resistance of a Medium and the Motion of Heavy Bodies Projected in a Resisting Medium], is a topic Leibniz had been working on since his time in Paris (where in early 1675 he had composed a tract, *Du frottement*, for the Paris *Académie des Sciences*), and which he had also discussed with Edme Mariotte during 1683 and 1684; as he notes, his results in the *Schediasma* are obtained by the use of his '*calculus of sums and differences*', the elements of which he had sketched in his 1684 article in the *Acta eruditorum*; but here Leibniz only presents his results in geometric language for the ease of comprehension of his contemporaries.[41] But he gives no analysis of the results Newton had published on motion in a resisting medium in Book II of his *Principia*. The third ([**10**]) and most substantial paper was Leibniz's presentation of a mathematical treatment of celestial motions based on a neo-Keplerian vortex theory, the *Tentamen*. This was his chief response to Newton's triumphant claim in his *Principia* to have solved the motion of the planets by his successful derivation of the inverse square law of gravitation and Kepler's three laws.

To a modern reader, Leibniz's publishing of these papers in the wake of the *Principia* is apt to seem puzzling. Why, if he had obtained these results through his calculus already some years earlier, had he not published them before? And what was the point of publishing them after Newton?

To answer these questions it might be useful to say something about publication strategy in the seventeenth century. Then, as now, one of the main purposes of publishing was to establish priority—if not of the results themselves, then certainly priority of the methods used to obtain them. But this could be achieved by other means, especially by sharing with those in one's circle. Newton was particularly loath to air his views in public, especially after his early papers in optics had been criticized by Hooke and Huygens. So, concerning his mathematical manuscripts in particular, he would share them with selected individuals whom he deemed amenable to instruction and who could be trusted to keep them secret—what Niccolò Guicciardini has called a 'policy of controlled scribal publication'.[42] Indeed, it took a considerable effort on the part of Edmond Halley to persuade him to publish his

[41] For a detailed analysis of this paper showing Leibniz's complex calculations using the calculus, see Aiton (1972a), and Bertoloni Meli (2006, 301–2).

[42] See Guicciardini (2009, esp. 339–65) for an account of Newton's policy on publishing.

mathematical treatment of planetary motion, a project that eventually resulted in the *Principia*.

Leibniz, too, often seemed to be in need of an external stimulus, as evidenced by his failing to publish his differential calculus until some nine years after its conception, in order to correct Tschirnhaus' account of the method of quadrature and to establish his priority.[43] But, in contrast to Newton, he was very keen on cooperation among scholars and on the establishing of scientific journals to that end.[44] He was pleased, rather than challenged, on discovering John Craig's use of his differentials in the latter's article of 1685, and in his reply the following year sought to show how more could be achieved by his methods than what he found in Craig's article. Subsequently, he was also pleased to find that the Bernoulli bothers, Jacob and Johann, had taken up his methods, and the three subsequently competed in solving problems by applications of Leibniz's calculus, as we shall see later in this chapter.

Of course, Leibniz had already published these articles of 1684 and 1686 containing the rules for his differential calculus by the time Newton composed the *Principia*, knowing that Newton was in possession of his own powerful methods. In two famous letters of 1676, in response to Leibniz's request, Newton had guardedly described some of his results in analysis (without revealing the method of quadrature on which they were based). Since that time, however, Newton had become increasingly disdainful of analysis. He regarded its methods as being valuable heuristic devices for the discovery of results, but as not meeting the standards of rigour and certainty achievable by geometric proofs in the style of the ancients—the paradigm of which had been provided by Huygens in his 1673 treatise *Horologium Oscillatorium*. Consequently he had formulated his masterwork almost entirely in geometrical terms—more accurately, in terms of his new synthetic geometry, in which geometric quantities ('fluents') are continuously generated in time, possessing at all times corresponding rates of change ('fluxions')—with (almost) all demonstrations given in the traditional geometrical fashion.[45]

43 See footnote 22 above for references.

44 See, for example, what he wrote in his reply to Jacob Bernoulli in 1691: 'I published some elements of this [sc. the calculus] a few years ago, being mindful of public utility rather than personal glory, and I might perhaps have been further on my way to the latter had I suppressed the method. But to me it is more pleasing to see fruit growing from the seeds I have scattered in other people's gardens too' (see **[18]** below, p. 439).

45 For a masterly account of Newton's (earlier) analytical and (later) synthetic methods of fluxions, see Guicciardini (2009, chs 8 and 9). Perhaps prompted by Leibniz's 1684 publication (1684b), Newton also gave a systematic rendition of his direct method in the Scholium to Lemma 2, Book 2 (§9.4), whereas in his treatises *De Analysi* and *De Methodis* he had illustrated it only through particular examples (Guicciardini 2009, 333). But in the *Principia* he still refrained from disclosing his analytical method of quadratures (362).

This accounts for one of Leibniz's probable motivations in publishing these papers after Newton. He wanted to show the power of his calculus in delivering analytic derivations of results Newton had achieved through his synthetic geometry. What was at stake was not the results themselves, but the power of the methods by which they had been derived. Here, though, Leibniz was somewhat handicapped by the fact that his contemporaries could not readily understand his new calculus, so he had to present the results in a geometric form that they would understand, while making clear that he had used his differential calculus to derive them.[46] It is only in the third paper on the celestial motions, the *Tentamen*, that he actually gives the differential equation from which he derives the equation governing the motions.

In the cases of the treatments of the optical lines and motion in a resisting medium, these papers were based on Leibniz's previous work, as he had claimed.[47] This was not the case, though, with the calculation in the *Tentamen*. It is certainly true that he had previously had 'some thoughts concerning the physical causes of celestial motions', since he had espoused the vortex theory popularized by Descartes as early as his *Hypothesis Physica Nova* of 1671. But his insinuation that he had only seen Pfautz's review of the *Principia* before composing his *Tentamen*, and had suppressed his solution so as 'to compare more carefully geometrical laws with the latest phenomena of the astronomers', has been thoroughly exploded by Bertoloni Meli's (1993) comprehensive study.[48] As Meli has shown, there is a clear trajectory from the notes and marginalia Leibniz had initially made on the *Principia*, through his first mathematical formulations of his own theory in his unpublished papers, to its final elaboration in the *Tentamen* on the basis of a physical hypothesis, proving that Leibniz already had access to Newton's *Principia* in Vienna in 1688, contrary to his claims to have worked simply from Pfautz's review before studying the book and making excerpts from it in Rome.

[46] Christiaan Huygens, for example, wrote to Leibniz in August 1690 that 'I have seen from time to time something of your new algebraic calculus in the *Acta* of Leipzig but, finding some obscurity in it, I have not studied it enough to understand it, and also because I believed I had something of an equivalent method, both for finding tangents to curves where the ordinary rules do not apply (or are very difficult to apply), and for several other investigations' (A III 4, N. 271, 547).

[47] For an account of Leibniz's earlier work in optics, as well as its later development, see McDonough (2018).

[48] See Leibniz's remark about this in the *Tentamen* (chapter 10). Also, in his (unpublished) *Illustratio Tentaminis De Motuum Coelestium Causis* [Illustration of the Essay on the Causes of Celestial Motions] of 1705 (GM VI 255), he wrote 'It is known that at the time when I sent the *Tentamen* to the editors of the *Leipzig Acta* I was on a journey through various places, and had not yet inspected the book of Newton's *Principia*, but had only seen a review of it made in the *Acta*, as I acknowledged there (*Tentamen*, §20) and now note again, because the most learned objector (in the said book 1, Prop. 77, p. 99) seemed surprised that I had put out such a *Tentamen* after the Newtonian *Principia* had been published.'

This dissimulation on Leibniz's part was a serious error of judgement, and one that cost him dearly in his relationship with Newton, contributing to the later acrimonious dispute over the origins of the calculus. To Newton it seemed clear that Leibniz had simply taken his result in Proposition 11 of Book 1 of the *Principia* and dressed it up in terminology of his own devising afterwards, allegedly making serious mathematical mistakes as he did so.[49] Moreover, from his perspective, even if Leibniz had been successful in deriving the inverse square law from his differential equation, this was pointless, since Newton regarded such a method as failing to deliver the certainty provided by the geometrical proofs he himself had given. Here Newton had the weight of tradition behind him. Even partisans of the new analysis who saw themselves as promoting and perhaps extending methods of discovery possessed by Archimedes, still acknowledged the necessity of geometrical constructions to establish demonstrations of those results.

Nevertheless, as is clearer to us in hindsight, this is to underestimate the power of the methods Leibniz had developed. Perhaps indicative of this underestimation by his contemporaries was the lack of response to the challenge problem he had set for scholars to solve in his second reply to Catelan, namely the determining of the isochronous line. By the time he revealed his own solution, published in the *Acta eruditorum* while he was in Rome in April 1689 ([**12**] in this selection), only Huygens had offered a solution. Leibniz's own solution, while agreeing with Huygens', was a wonderful testament to the power of his calculus. And, lest the Cartesians might feel that they had been pre-empted by Huygens' solution, Leibniz proposed a further problem for them to solve, that of finding the curve traced by a body equably approaching or receding from a given point under its own weight.[50]

Although Leibniz had clearly worked with Newton's geometrical constructions in order to set up his differential equation in the *Tentamen*, to regard Leibniz's theory as simply recapitulating Newton's results is a disservice to his mathematical accomplishment, and also ignores the fundamental difference in their philosophies of how mathematics should be applied to nature. For the derivation of his differential equation was no simple matter, and required considerable skill in performing changes of variables and integration. It

[49] For an analysis of Newton's criticisms, see Aiton's masterly account in his (1972b, 136–45), where he defends the validity of Leibniz's use of differentials against Newton's later objections.

[50] The resulting curve is the 'paracentric isochrone'. A solution to this problem was first published by Jacob Bernoulli in the *Acta eruditorum* of June and September 1694, to be followed by Leibniz's solution in the August issue and Johann Bernoulli's in the October issue of that year. See O'Hara's introduction in his (forthcoming) for discussion of the other challenge problems that followed the ones reproduced here in the years 1692–96.

depended on Leibniz's recognition that Newton's resolution of the nascent motion of the orbiting body into an attraction towards the centre and an inertial motion along the tangent is not unique: he discovered that the same nascent motion could also be resolved into a circular motion of the radius and a motion of the orbiting body along the radius, either towards or away from the centre. The first of these he called a *harmonic motion*; the second, a *paracentric* motion. The latter came out beautifully in his solution of his differential equation for elliptical motion, according to which the 'paracentric solicitation' towards the centre is given by a two-term expression: $ddr = aa\theta\theta{:}r^3 - 2a\theta\theta{:}r^2$, where $a\theta$ is the element of time dt. Leibniz interpreted the first term as representing the centrifugal force on the body away from the centre, and the second as representing the gravitational tendency on the rotating body, given by an inverse square law. This, of course, is highly reminiscent of Kepler's theory in his *Astronomia nova* of 1609, according to which an orbiting planet is pushed around in its elliptical orbit by the rotating rays arising from the sun's magnetic force.[51] This circulation happens, according to Kepler, 'after the manner of an impetuous vortex' by which the planets are borne in a circular motion, but 'with a stronger or weaker thrust according as, by the law of its emanation, it is denser or rarer' (Aiton 1972b, 13). It only remained for Leibniz to reformulate the idea in terms of hypotheses acceptable to the mechanical philosophy: instead of the planets being pushed around by the rays of an immaterial virtue, they were immersed in a material vortex, just as Descartes had proposed, but in such a way that the component of their velocity perpendicular to the radius vector is inversely proportional to their distance from the sun, as Kepler had proved in his *Epitome*. The latter was what Leibniz now termed *harmonic motion*. Kepler had given a second cause for the motion along the radius vector (Leibniz's *paracentric motion*), as due to 'another solar virtue distinct from the first, a kind of magnetic attraction and repulsion' (Kepler 1937, vii, 377; Aiton 1972b, 17). On Leibniz's model the motion along the radius was due to an imbalance between the centrifugal force away from the centre and a tendency towards the centre inversely proportional to the square of the radius, resulting in the elliptical orbit. Thus Leibniz's mathematical resolution allowed him to satisfy at one stroke both Kepler's demand that astronomy be based on physical, not merely mathematical hypotheses, and the essential requirement of the mechanical philosophy, that all action be

[51] '*Flumen est species immateriata virtutis in Sole magneticæ* [The flow is a non-material species of magnetic virtue in the Sun]' (Kepler, *Astronomia Nova* (1609) ch. 57, p. 270; Aiton 1972b, 26).

by contact.[52] And on that basis he had given a derivation of the inverse square law of gravitation from Kepler's first two laws—although without a mechanical explanation of Kepler's second solar virtue.

This explains a further main motivation Leibniz had for publishing his theory as an alternative to Newton's. For it was not Leibniz's but Newton's contribution that was the outlier in the theory of celestial motions at the time.[53] Despite the technical and mathematical brilliance of the *Principia*, acknowledged by all who had managed to gain access to it, it was not until many years later that Newton was accredited with having given a successful account of gravity—in France, vortex theories were still being proposed as late as 1760, where a mechanical explanation of gravity was still being sought.[54] The reason for this is simple: a successful scientific explanation, according to the dominant paradigm, was one that provided the cause of the phenomenon under investigation.[55] Newton had not provided a physical cause for his universal gravitation; nor did his theory explain the fact that the planets all rotate around the sun in the same plane, the plane of the ecliptic—although of course, the comets did not, as the Newtonians were at pains to point out.[56] Thus Huygens expressed a representative reaction to the *Principia* when, after expressing his great esteem for Newton's knowledge and subtlety, he declared that 'he has in my opinion made a poor use of them in most of this work, when

[52] See Leibniz's explanation in the *Tentamen*: '*all bodies that describe a curved line in a fluid are set in motion by the fluid itself*'.

[53] Leibniz's model could, for example, be compared with that of Giovanni Alfonso Borelli, where the planets were pushed around in their orbits by corporeal rays emanating from the rotating sun, as if by a lever rotating with it. For him the inward tendency was due to an appetite of the planets, the outward one due to the centrifugal force. See the accounts of Borelli's theories given in Aiton (1972b, 91–4) and in Bertoloni Meli (2006, 193–202).

[54] Newton himself had sought a mechanical explanation of gravity in terms of an aether, first in his 'An Hypothesis explaining the properties of light discoursed of in my severall papers', then again in the *Opticks* of 1704. See Aiton (1972b, 106–9). In between, the Swiss mathematician Nicolas Fatio de Duillier (1664–1753) had written to Leibniz that he had a mechanical hypothesis that he could 'demonstrate was sufficient for him to explain precisely all the phenomena of gravity', that he was saying 'nothing he could not prove', and that 'Mr. Newton agrees with the exactness of my demonstrations' (Fatio to Leibniz, 9 April 1694; A III 3, addenda).

[55] We might add that the derivation of Kepler's Laws was a matter of secondary importance to the astronomers, who were by and large content to regard them as empirically established. Natural philosophers, for their part, were often surprisingly indifferent to Kepler's Laws, regarding Descartes as having provided a satisfactory physical explanation. Malebranche, for example, completely ignored them, and appears to have been unaware of the third law until he read about it in Huygens's *Cosmotheoros*, published posthumously in 1698. See Aiton (1972b, 72).

[56] Thus when Huygens criticized his theory for appealing to Cartesian vortices while Newton's explained the motions using inertial and centripetal motion alone, Leibniz responded: 'And the reason I still do not repent the deferent matter after having learned of Mr. Newton's explanation, is, among others, that I see all the planets go in the same direction, and in the same region [*sc.* in the same plane], which can also be remarked of the little planets of Jupiter and of Saturn. Instead of which, without the deferent matter in common, there is nothing to prevent the planets form going in every direction' (to Huygens 16/26 September 1692, GM II 144).

the author researches things of little utility, or builds on the scarcely plausible principle of attraction'.[57]

For Huygens had his own theory of gravitation, based on Descartes's vortex theory together with his own research on centrifugal force. According to Descartes, the tendency of bodies moving in a circle to continue their motion in a straight line tangent to the circle would result in a tendency to move outwards from the centre. This tendency would then have to be counterbalanced by a tendency of the surrounding fluid in order to keep them from straying from their circular motion, as seen in the vortical motion of fluids. Huygens improved on this analysis, showing how the tendency outwards (for which he had coined the name 'centrifugal force') would be proportional to the square of the orbital speed for any given mass, and the motion a simple composition of the tendency along the tangent and the centrifugal one. In this way he was able to give quantitative form to Descartes's vortex theory.

All this was supported by ingenious experiments. One of these is described by Denis Papin in his 'Observations on the Cause and Properties of Gravity', [**11**] in our collection, appearing in the *Acta eruditorum* of April 1689. Although this was not an experiment that he had helped Huygens set up while he was acting as his assistant in Paris, Papin enlisted it in the process of supporting Huygens' theory of gravity. In the experiment, two identical balls are attached by a piece of string, one suspended under a rotatable board through a hole in its middle, the other rotating with the board. By this means Huygens was able to determine how fast the rotation would have to be for the centrifugal force on the rotating ball to counterbalance the weight of the suspended one. He concluded that if the Earth turned on its axis seventeen times faster than it actually does, then the centrifugal force on a body at the equator would equal the weight of the body there.[58] If, then, a vortex of aether circulated the Earth many times faster than the Earth rotated, so that the centrifugal force on its particles pushed them outwards, they would displace the body inwards, thus explaining its weight by the action of the aether, in accordance with Descartes's theory. But, of course, this would only give a tendency towards the Earth's axis, and would also decrease to zero at the poles, as had been objected by Johann Christoph Sturm and Jacob Bernoulli in an

[57] OCH, x, 354; translated from the French quotation given by Guicciardini (2009, 326, n. 53).

[58] This provided evidence for the reality of the Earth's rotation. For since the centrifugal force on the body would be proportional to the square of its speed, the apparent weight of an object at the equator should be reduced by a factor of $1/17^2$ or 1/289 compared to its weight at the poles, with intermediate values in the latitudes between; as Huygens was able to confirm experimentally by comparison of measurements of the weight of a body at and Paris. See Aiton (1972b, 83–4).

article in the *Acta eruditorum* of February 1686.[59] So Huygens had proposed instead an aether circulating in great circles all around the Earth, resulting in an equal distribution of the upwards tendency of the aether, and thus of the downwards tendency of ordinary bodies.

Huygens had presented his theory in a memoir to the Paris *Académie* in 1669, and it had since been commented on by many scholars there, including Roberval, Mariotte, Claude Perrault, and Pierre Sylvain Régis.[60] But he did not finally publish it until 1690, as a supplement to his *Traité de la Lumière*, titled *Discours sur la cause de la pesanteur*. By this time, of course, he had studied Newton's *Principia*, as well as Leibniz's *Tentamen*; we will come back to his reactions to their theories below.

First, though, let us return to Papin's article of April 1689 ([**11**]), in which he claimed that Huygens's theory of gravity already contained the resources to defend the Cartesian vortex theory from the criticisms of Sturm and Jacob Bernoulli, and that it also revealed the source of Leibniz's errors in criticizing the Cartesian measure of force. He begins by asserting a proposition he takes Huygens to have demonstrated, namely that the aether Huygens proposed as the cause of gravity '*has a speed that is infinite in comparison with the velocities of the falling bodies that may be observed by us*'. On the basis of the experiment described above, he notes, Huygens had calculated that the aether causing gravity would have to circulate so fast that 'it would be able to traverse the whole circumference of the Earth almost a thousand times in any single hour'.[61] This would mean that the differences in the force of gravity at different heights would be comparatively negligible. Consequently, bodies could still be interpreted as receiving equal quantities of motion from the impacts of the aether particles in equal times, in accordance with Galileo's law of fall. From this fact Papin inferred that the objections to the Cartesian vortex theory advanced by Sturm and Bernoulli were easily countered, as the difference in velocities of bodies at different latitudes would be negligible compared to the speed of the aether causing gravity. Likewise, Leibniz's

[59] '*Dn. Bernoulli Dubium circa Causam Gravitatis a rotatione Vorticis terreni petitam* [Mr. Bernoulli: A Doubt concerning the proposed Cause of Gravity by the rotation of a terrestrial Vortex]', *Acta eruditorum*, February 1686, 92–5.

[60] Gilles Personne de Roberval (1602–1675) was a noted French mathematician and rival of Descartes, who independently invented (but never published) a method of indivisibles akin to Cavalieri's. Roberval also published what was purportedly a Latin version of a lost work of Aristarchus, but was in fact his own affirmation of the Copernican hypothesis in which he proposed a mutual attraction among all particles of matter. Claude Perrault (1613–1688) was a French physician, architect, and natural philosopher, and Pierre Sylvain Régis (1632–1707) was a Cartesian philosopher and prominent critic of Spinoza.

[61] In his second reply Papin had to retract this claim, ceding that he had vastly overstated the speed Huygens had calculated, having cited it from memory.

objections to Descartes's concept of force as quantity of motion depended on his assumption that motive forces are proportional to the space fallen through by a falling body. But if a falling body receives equal quantities of motion from the aether in any given time, no matter how far it has fallen, then the motive forces are proportional to times, not to the space fallen through. Likewise the resistance to motion experienced by an ascending body will be proportional to the time, not the space traversed, and in equal times forces will be diminished by equal amounts. So Leibniz's assumption that the force of gravity on a falling body is proportional to the space traversed cannot be accepted.

Papin's article containing this critique came out in April 1689, when Leibniz was in Italy, and he did not find it until he was on his way home to Hanover in April 1690. He had made substantial progress in physics in Italy, composing several new manuscripts. One of these was the *Tentamen de physicis motuum coelestium rationis*, which remained unpublished. A second manuscript that suffered the same fate was a revised version of the original *Tentamen* of the *Acta eruditorum*, the so-called '*zweite Bearbeitung*', which was probably composed in stages between the summers of 1689 and 1690. This was evidently intended for publication in Italy, since in it Leibniz removes the favourable introductory references to Kepler and Copernicus, presumably in an effort to avoid the censors.[62] In Rome and in Venice he had been involved in efforts along with Galileo's disciple Viviani, by appeal to the censor Fr. Baldigiani, to have the prohibition of Copernicanism revoked, and had drafted a number of manuscripts on Copernicanism and the relativity of motion in an effort to support this.[63] More substantively, in the *zweite Bearbeitung* Leibniz returned to the explanation of gravity, positing (like Kepler) a different cause for gravity from that responsible for moving the planets. While the planets were carried in a vortex by a fluid consisting of coarse particles, the aether responsible for gravity consisted of much finer particles, of the kind supposed to be responsible for magnetism.

During his stay in Rome Leibniz had also been busy developing his new science of dynamics. His first try was a dialogue in two parts, the *Phoranomus seu de potentia et legibus naturae*, based on his conversations with the

[62] In February 1690 Bodenhausen had written to Leibniz advising him that if he wished to publish a version of the *Tentamen* in countries under the control of the Inquisition he should omit the favourable mentions there of Kepler, Galileo, and Copernicus, 'and several expressions which will arouse suspicion in relation to the motion of the Earth' (A III, 4 469; quoted from O'Hara, forthcoming, ch. 2). (Leibniz had left the manuscript with Bodenhausen; on whom, see footnote 65 below.)

[63] See C 590–3/AG 90–4, and A VI 4c, 2065–7, 2068–9, and 2070–5, translated in Arthur (2021, Appendix 4).

members of the Accademia Fisico-Matematica.[64] But he broke off work on this in the summer of 1689 to begin composing a major work, the *Dynamica de potentia et legibus naturæ corporeæ*. This was an attempt at a synthetic account of physics from its foundations upwards, perhaps conceived as a counterpart to Newton's *Principia*. He was not able to finish it in Italy, leaving a draft manuscript of it in the hands of his friend Bodenhausen in Florence for redaction and proposed eventual publication.[65] But this was never achieved; to Bodenhausen's mounting frustration, Leibniz sent revisions, then pleaded the need to include new considerations, which he was never able to complete to his satisfaction.

On stopping off in Vienna in March 1690 on his way home to Hanover, Leibniz discovered Papin's article in the April 1689 *Acta eruditorum*, to which he sent off a reply to Otto Mencke the following month ([**13**] in this collection). When that came out in the May 1690 *Acta eruditorum* he discovered two articles by Jacob Bernoulli in the same issue that were deserving of response. The first concerned a solution to a dice game problem Bernoulli had himself set in 1685; the second presented a new solution to Leibniz's challenge problem of the isochronous curve which—Leibniz was delighted to discover—was solved using the calculus whose elements he had published back in 1684.[66] So Leibniz immediately sent off a reply to Bernoulli's articles that was published in the July 1690 issue ([**14**] in this collection), showing that solutions to the first problem yielded infinite series that were not at that time solvable. In his second article Bernoulli had proposed a further problem that could perhaps be solved using the same methods of the differential calculus, namely determining the catenary line, or hanging chain, first described by Galileo. Leibniz took up this suggestion and promoted it into a challenge problem, for which he proposed contestants should submit their solutions by the end of the year.

The reply to Papin ([**13**]) is a model of clarity, obviating the need for lengthy exposition. As Leibniz explains, Papin had misunderstood the point of Sturm's and Bernoulli's criticism of Descartes's theory, which had nothing to do with the question of whether there was any variation of gravitational force with height, but rather addressed the fact that on Descartes's account the force

[64] For discussion of the *Phoranomus*, see Garber and Tho (2018, 316–20), Duchesneau (1994, 151–73).

[65] Baron Rudolf Christian von Bodenhausen (1640–1698) was a German nobleman who served as tutor to the sons of Grand Duke Cosimo III of Florence. Leibniz had met Bodenhausen in Florence, and had discussed his calculus and new science of dynamics with him.

[66] It was in this article that Bernoulli first used the term 'integral' instead of Leibniz's term 'sum'.

would be directed to the axis of revolution of the Earth. Huygens had not been offering to rescue Descartes's vortex theory, but was presenting a different hypothesis of his own. Moreover, Leibniz added, a proper explanation of why variations in the strength of gravity could be ignored (at any rate for distances small in relation to sense) would be given in his forthcoming *Dynamics*—the first reference in print to the existence of this work. He then gives a very clear explanation of Huygens' hypothesis of gravity being caused by the centrifugal force of an assumed circulating aether. He notes that the difficulty about its tending towards the axis of rotation is avoided by supposing that the aether 'move around the Earth not along the equator or in parallels, but in great circles like the meridians', but argues that still in this case there would be a concentration of aether towards the poles. Instead he tentatively suggests another hypothesis, that a certain matter 'explodes' outwards from the Earth in all directions, producing a kind of radiation like light, effectively pushing down the grosser bodies towards the centre.[67]

Before turning to Leibniz's responses to Papin's other criticisms and the controversy over the Cartesian measure of force, let us stay with this question of the cause of gravity. For while Leibniz was returning from Italy, Huygens had published his *Discours sur la cause de la pesanteur* in early 1690, and in February of that year he had written to Leibniz enclosing a copy of the *Traité de Lumière* containing it. Unfortunately, however, Leibniz did not receive this package until almost October of that year, since van der Heck, to whom Huygens had entrusted it, was unable to locate the itinerant Leibniz. In the meantime Huygens had also been travelling, and had visited Newton in person on a trip to England in 1689, accompanied by Fatio de Duillier, who, while living in England, had struck up an intimate friendship with the reclusive English savant. As a result of his discussions with Newton and a close reading of the *Principia*, Huygens found himself very impressed with Newton's mathematical results. In particular, despite rejecting the idea of a universal attraction of all particles to one another, he accepted Newton's geometrical proof of the inverse square law of the gravity of the planets towards the Sun, and his refutation of the vortex theory. So in his letter of February 1690 he asked Leibniz whether he too had changed his mind on planetary theory after having

[67] Leibniz reveals this theory to Huygens in a letter of 26 April 1694. In response to what he has learned of Fatio de Duillier's theory (see footnote 54), Leibniz tells Huygens that 'in fact I already imagined on other occasions that there could be a type of explosion or *recessus*, the ejection of a very fine, and consequently a more solid or, if you like, denser matter, that would consequently oblige the rarer and coarser matter to approach [the emitting body]' (GM II 171). See also his letter to Des Billettes, 4/14 December1696 (GP VII 452/L 473) for an explanation.

seen Newton's book.[68] He could not, for instance, see the necessity of Leibniz's assumption of harmonic motion, since Newton had derived the inverse square law without it, supposing only a centripetal tendency balancing the centrifugal tendency of the orbiting planets.[69] Moreover, in his *Tentamen* Leibniz had not actually offered any hypothesis as to the cause of gravity, acknowledging this as one of two matters he had left unexplained (the other being Kepler's Third Law).

Leibniz drafted a reply addressing these objections once he had received Huygens' letter, although he subsequently decided not to send it. He reiterated his acceptance of the inverse square law for the apparent attraction of gravity, and the need to find an acceptable mechanical explanation, like that of Huygens. He also argued that he could account for the stability of vortices on the hypothesis that 'there is the same quantity of power in each orbit', making clear that by 'power [*puissance*]' he meant his measure of force in terms of quantity of the effect 'from which it follows that the absolute forces are as the squares of the speeds'. On this basis he was able to derive Kepler's Third Law, that 'the squares of the periodic times are as the cubes of the distances'.[70] Although, as he himself admitted, the hypothesis was not compatible with the hypothesis of harmonic circulation,[71] that difficulty could be overcome if one supposed (as in the *zweite Bearbeitung*) a different cause for gravity than for circulation, in this case an aether causing gravity different from (and finer than) the aether carrying the planets. In the reply he actually sent to Huygens, however, Leibniz did not mention the theory of planetary motions at all.

[68] 'But since in treating this matter you had still seen only an extract of his book and not the book itself, I would really like to know whether you have since changed your Theory, because you are bringing in the Vortices of Mr. Descartes, which in my opinion are superfluous if you accept Mr. Newton's System, where the motion of the planets is explained by the gravity towards the Sun and the centrifugal force, which counterbalance one another.' Huygens' letter is dated 9 February 1690 (OCH, ix, 366; GM II 41).

[69] Here we are in complete agreement with the analysis of Bertoloni Meli (1993, esp. 197–8), according to which Huygens, Leibniz, and Newton all regarded centrifugal force as a real force. While it is now understood that centrifugal acceleration is an effect produced in a rotating frame of reference, and that in a stationary reference frame there is only a centripetal force continuously deflecting the body towards the centre (as in the resolutions of curvilinear motion adopted by all three men), these thinkers nevertheless persisted in conceiving the situation as involving an equal and opposite reaction between centrifugal and centripetal forces.

[70] Quotations from Leibniz's (unsent) October letter to Huygens are from (OCH, ix, 525–6; Aiton 1972b, 134). As Bertoloni Meli has shown, the descriptions in this letter match well with another manuscript Leibniz had composed, the *De causis motuum coelestium*, which therefore probably also dates from October 1690. See Bertoloni Meli (1993, 165–8).

[71] This is because on this hypothesis $v_1/v_2 = (r_2/r_1)^{1/2}$ for each circular orb, whereas harmonic circulation required $v_1/v_2 = (r_2/r_1)$.

Meanwhile Papin had drafted a reply to Leibniz, but it did not appear until the following year in the January issue of the *Acta eruditorum* of 1691 (article [**15**] in this collection). In it he appealed to Huygens' *Discours* in defence of his own description of Huygens' views on gravity. In the version published in the *Traité de Lumière*, Huygens himself had carefully separated the original of his *Discours* from comments he made after seeing Newton's *Principia.* Papin, though, makes no reference to the latter, and proceeds as though Huygens was still fully committed to a vortex theory, simply referring the reader to Huygens' book for a response to Leibniz's objections. He then returns to the controversy over whether the quantity of motive force is correctly estimated by the quantity of motion.

In his article of the previous May ([**13**]), Leibniz had proceeded systematically, explaining his views on the quantity of force through careful definitions and a thought experiment designed to appeal to Papin's practical mechanical turn of mind. Not wanting to be bogged down in definitional disputes about force, Leibniz had tried to get Papin's acceptance of the principle by which he claimed to have discovered the true measure of force, his Equipollence Principle, according to which 'the power of the full cause is the same as that of the entire effect'.[72] His idea was that if these powers were unequal, then perpetual motion could be obtained, a consequence he took to be so obviously untenable that a reduction to such a consequence would have the same force in mechanics as a reduction to absurdity in logic. He then imagines an apparatus whereby the motive force acquired by a body *A* through falling through a certain height is wholly imparted to a body *B* at rest on a horizontal plane. *B* then moves up along an inclined plane to whatever height its force will take it. Now if *A*, by falling through a height of 1 unit, has a mass of 4 and a degree of speed of 1, then according to the Cartesian measure of force, a body *B* of mass 1 will have a motive force of 4, and thus a degree of speed 4. But now according to the Galilean formula, the height through which it can be raised by this force (assuming no friction) will be proportional to the square of the degree of speed, namely 16 units. This being so, by falling, say, onto the end of a balance 4 times as far from the fulcrum as the original globe *A* at the other end, and thereby imparting all its force to *A* of 4 units, it could raise *A* to a height of 4 units, since 16 divided by 4 equals 4. But now *A* has four times as much force as it did at the beginning, and perpetual motion could be produced.

[72] Leibniz first employed the Equipollence Principle for the purposes of the eventual dynamics in his *De arcanis motus et mechanica ad puram geometriam reducenda* of 1676: see Duchesneau (1994, 103 ff.) and Garber and Tho (2018, 307–14) for discussion.

One might object that here Leibniz has mixed a dynamical measure of force with one drawn from statics. The force accrued in rising 16 units should be proportional to mv^2, but the force imparted to the end of the lever through its fall will be a force of impact, proportional to mv. In his reply Papin ([**15**]) does not challenge him on this, however, precisely because he is in any case interpreting force as directly proportional to speed. Instead, he challenges the terms on which Leibniz's thought experiment was founded. Although he accepts Leibniz's proposal that the force or power of a body be measured by its effect, he objects that this effect is not to be estimated by the space traversed or time taken, but by 'the resistance that it overcomes'. He then argues that a body ascending four times as fast up an inclined plane would experience four times as many impressions of gravity that it would have to resist, with the number of impressions exactly compensating the strength of resistance that a body four times heavier would have to counteract in climbing the same height. Thus, a four pound body ascending an inclined plane would experience exactly the same resistance as a one pound body with four times its speed ascending an equivalent inclined plane. Therefore, their forces would be equal. As for Leibniz's argument, Papin simply denies his assumption that the whole power of the body *A* could be transferred into the body *B*.

Leibniz's response (in the September 1691 issue of the *Acta*, [**19**] in our collection) is again very clear. He neatly reduces Papin's argument to two syllogisms, one (the prosyllogism) given in justification of the minor premise of the other, and explains how the prosyllogism assumes as a premise what the whole argument is supposed to be establishing, namely that the force (here equated with the resistance to the falling body) is 'to be estimated by the product of velocity and quantity of body, that is to say, quantity of motion'. Here Leibniz anticipates the idea of potential energy, arguing that for him 'those things are *equal in force* which can bring an equal number of springs equal in force to the same degree of tension, or which can raise the same number of weights to the same height above the prior situation of each of them'. Motion, for him, is only a mode of body, whereas force is a power a body can retain even when it is not in motion. And when this power is used up in setting the body in motion, he takes himself to have shown, this power is not proportional to speed.

The key to Leibniz's estimation of force is that it should be a measure of the power to achieve an effect which is *conserved under repeated substitution*: it is a quantity preserving that power that remains intact in the effect, that is, under substitution of effect for cause. This echoes the methodology of his logic, where truth is preserved by the substitution of identical propositions for one another (substitution of identicals *salva veritate*); but it also tallies with the methodology of Leibniz's mathematics, where equal things are those that can be substituted for one another *salva quantitate* and similar ones *salva qualitate*.[73] Leibniz is disappointed that he cannot convince Papin of this axiom; but he is not discouraged, and suggests they continue to try to resolve their differences in private correspondence. This news is welcomed by the editor of the *Acta eruditorum* Otto Mencke, who asks them, once they have resolved their differences, to convey a synopsis of the controversy for publication in his journal. Unfortunately, though, despite the best efforts of the two men, who exchanged 16 letters between January and December 1692, including a draft synopsis by Papin, they were unable to come to any agreement, with Papin refusing to accept that the total force of one body could be transferred to another, even in principle.

Between reading Papin's second reply of January 1691 ([**15**]) and sending off his response in September ([**19**]), Leibniz had been busy with other matters. One of these was to take up again the question of motion in a resisting medium. Huygens had published his own results on this topic (which he had worked out long before)[74] in the *Discours sur la cause de la pesanteur*, of which (as we have seen) he had sent Leibniz a copy early in 1690. In his letter of August 1690 Huygens prompted Leibniz for a response on the topic, and also raised an objection to Proposition 6 of Leibniz's account of the resistance of a medium in 1689 concerning the composition of motions ([**9**]).[75] Leibniz accepted the criticism, and after further exchanges with Huygens, decided to publish a correction, using the opportunity to explain the relation between his own results and those of Huygens and Newton. This comprises [**16**] in our collection.

Moreover, in July 1690 Leibniz had requested solutions to Jacob Bernoulli's challenge problem of the catenary by the end of that year, a deadline that had now passed, and two solutions had appeared: one by Huygens and one by Jacob Bernoulli's younger brother, Johann. Leibniz sent off his own solution ([**17**]) to the *Acta eruditorum*, which was published along with those of Huygens and Johann Bernoulli in the issue of June 1691, with a new solution from Jacob Bernoulli appearing immediately afterwards.[76] This was an important moment for Leibniz in demonstrating the advantages of the

[73] See David Rabouin (2013).

[74] Huygens' work on this dates from 1668 and 1669, and he had presented some of his results to the *Académie des Sciences* while he was in Paris.

[75] Huygens to Leibniz, 9 February 1690; (OCH, Vol. ix, p. 366).

[76] Tschirnhaus had failed to respond to the challenge problem, despite Leibniz's entreaties.

methods he was promulgating, since the form of the curve is a hyperbolic cosine, a transcendental curve. Such curves were regarded as untreatable by analysis, but Leibniz not only derived its analytic form using his calculus, he used this to show its geometrical construction in relation to an auxiliary logarithmic curve. Moreover, he derived expressions for the tangent, and also for the length of the curve between two points (showing that it is *rectifiable*), as well as the area under the curve (its *quadrature*), both of which had been regarded as impossible.

In September of the same year Leibniz published the comparison he had promised in the June issue of the three solutions to the catenary problem, his own, Huygens' and Johann Bernoulli's—article [**18**] in our collection. In it he notes how they all arrived at the same results through different routes, and explains the correlations among the different methods, stressing that his reduction of the matter to logarithms gives the most natural means of expression for transcendental curves, and also allows the clearest geometric constructions. He also notes how Jacob Bernoulli had since shown the illuminating connection between this problem and that of finding the loxodromic curve used in navigation, on which Leibniz himself had published in April.[77] Leibniz concludes the paper with a plea to Jacob Bernoulli for his opinion on Leibniz's first reply to Papin. Obviously this was on his mind, as he was sending off his second reply to Papin ([**19**]) at this time.

Meanwhile, Leibniz's correspondents were reminding him of his promise to put out a work on his new science, the *Dynamica*. In France his potential allies in his dispute with the Cartesians included Paul Pellisson-Fontanier, Gilles Des Billettes, Simon Foucher, Pierre Varignon, and Guillaume François, Marquis de l'Hôpital, and he needed to equip them with at least a sample to secure their support. In particular, he wanted to win Nicolas Malebranche over to his side, and to this end in January 1692 he sent an essay to Pellisson for the Paris *Académie des Sciences*, titled *Essay de dynamique*, or *Élemens de dynamique*, expecting publication in the *Journal des sçavans*. Instead, though, it was simply read to members of the *Académie* in July, and no copy was sent to Malebranche.[78] As he explained to Pellisson, in the *Essay* he had merely

[77] Bernoulli had discovered that the loxodrome, when stereographically projected, generates a *spira mirabilis*, or logarithmic spiral, which was the main topic of his paper, and noted the connection with the catenary problem. See McDonough (2022) for a discussion of the catenary as an example of optimal form, and for a stimulating account of the wider philosophical significance of this solution, as well as of Leibniz's and the Bernoulli brothers' later solutions to the brachistochrone problem posed by Johann Bernoulli.

[78] The manuscript remained in the archives of the *Académie des Sciences* in Paris until 1859, when Foucher de Careil included it in the first of seven volumes of Leibniz's works (*Œuvres de Leibniz* vol. 1,

extracted from the *Dynamica* what seemed to him best suited to explain the conservation of absolute force. Although the *Essay* begins with definitions, axioms, postulates, and propositions, its argument is essentially the same as the one contained in Leibniz's replies to Papin, where the Cartesian measure of force is shown to issue in perpetual mechanical motion.

The next article in our collection, the 'General Rule for the Composition of Motions' ([**20**]), is part of that same effort to win over scholars in France, and this time it did appear in the *Journal des sçavans* in September 1693.[79] In it, the motions that are composed are in fact quantities of motion, so that the mass of the body has to be taken into consideration in their composition. Accordingly, he explains how to compose the various tendencies to motion existing in a given body at a time into a resulting overall tendency, and also shows how quantity of progress (quantity of motion in a given direction) can be resolved and added vectorially.[80] The tendencies would be represented by differentials of his calculus, rendering a large suite of problems in applied mathematics tractable.

The article immediately following ([**21**]) presents two examples of this. The first was suggested to him by Tschirnhaus' treatment in his *Medicina mentis* [Medicine of the Mind] (1687) of the problem of determining the tangent of a curved line traced by a stylus pushing against one or several tensed threads. He notes also that Fatio de Duillier had solved the problem by a different 'very beautiful' means, and that the Marquis de l'Hôpital had solved it using his calculus. In fact, it was l'Hôpital's publication of his solution that had prompted Leibniz to publish this article in order not to fall behind his supporters in applications of the calculus. The second problem depended on his analysis of the centrifugal tendency, or the action of gravity balancing it, as composed of an infinite aggregate of solicitations, as described in his

470–83). Leibniz did manage to have an exchange with Malebranche in 1692, however, and apparently made some emendations to the text of his *Essay* soon afterwards. See O'Hara (forthcoming, ch. 3). This essay must not be confused with a second essay of the same title, *Essay de dynamique*, which Leibniz wrote in about 1701–02 (*Essay de dynamique sur les loix du mouvement, ou il est monstré, qu'il ne se conserve pas la même quantité de mouvement, mais la même force absolue, ou bien la même quantité de l action mortice* GM VI 215–31).

[79] See Pierre Costabel's erudite analysis of the 1692 *Essay de dynamique* in his (1960), in which he details its relationship with this paper on the composition of motions.

[80] This resolution of a motion into its orthogonal components is put to good effect by Leibniz in his correspondence with Johann Bernoulli. Building on an argument of Bernoulli's, Leibniz imagines a billiard ball *A* striking two other stationary balls *B* and *C* at once in such a way that they shoot off at right angles to one another (all the balls being of the same mass), with the cue ball stopping dead after transferring all its force to the other two. The velocities of *B* and *C* will equal the orthogonal components of the original velocity of *A*, which will be as the sides of a square to the diagonal, so that by Pythagoras's Theorem, $mv_A^2 = mv_B^2 + mv_C^2$ (Letter to Bernoulli, 28 January 1698; GM III-1 240), thus corroborating his conservation law for *vis viva*. See Lodge's account (Leibniz 2013, xlii–xliii).

Tentamen of 1689. The direction of the composite tendency is given by the line from the centre of rotation to the centre of gravity of the infinitely many solicitations, and its magnitude given by their sum, that is, integral.

Much of the resistance to his physics that Leibniz was receiving was due to a distaste for metaphysics on the part of his mathematically inclined correspondents. Leibniz had long believed that this was due to faulty metaphysics rather than any defect in the science itself, provided it were put on firm foundations. He was concerned in particular by the trend of post-Cartesian thinkers like Mariotte and Huygens to appeal to merely nominal empirical laws like the conservation of quantity of progress, without concerning themselves with either their foundation or their implications. It was a long-standing criticism of his that the notions of impenetrability and inertial mass they took for granted could not be explained using the purely geometrical concepts of the Cartesians, and also that if the quantity of force were equated with the quantity of motion, it would diminish in every inelastic collision, and therefore in the universe as a whole. On his understanding, metaphysical laws such as the Principle of Equipollence and the Law of Continuity could be appealed to, not only in weeding out theories that violated these principles, but also as a guide to constructing viable theories and attaining the correct interrelationships between the equations expressing the nominal laws that would conform to these principles.[81] In March 1694, Leibniz published an essay in the *Acta eruditorum* on this topic of the relationship of metaphysics to physics, '*De Primae Philosophiae Emendatione, et de Notione Substantiae emendatione* [On the Emendation of First Philosophy and the Concept of Substance]', promising that his new science of Dynamics 'would bring the strongest light to bear on understanding the true *concept of substance*', by basing it on an emended conception of active force.[82]

Needless to say, that only increased the pressure on him to give a fuller explanation of his dynamics, especially with his *Dynamica de potentia* no nearer to seeing the light of day. This Leibniz finally addressed by publishing the *Specimen dynamicum* in the *Acta eruditorum* of April 1695 ([22]). There, he explains that the force he conceives to be everywhere in things—to 'constitute the innermost nature of bodies'—is not a mere capacity, like the Schoolmen's 'active faculty', but is always manifested as 'an endeavour ôr striving, one that will have its full effect unless it is impeded by a contrary

[81] See Duchesneau (1994) for a thorough account of Leibniz's employment of these principles in founding and developing his dynamics.

[82] (GP IV 469/L 433). This essay will appear in the companion volume, *G. W. Leibniz: Journal Articles on Philosophy*, edited by Paul Lodge (Oxford University Press, forthcoming).

endeavour'. The connection with metaphysics is that it is of the nature of substances to act, so that the substance of bodies, far from being something merely passive like the Cartesians' extension, must consist in a force that God produced in bodies at Creation, and subsequently conserves in them. This primitive force that is responsible for the activity of bodies is only manifested through their derivative forces, with these arising from the conflict of bodies with one another. So, as Leibniz explains, the primitive power—a reinterpretation of the Scholastic substantial form, in the sense of a principle of activity—plays no role in explanations in physics, and the effects of the derivative forces are all such as to produce local motions, in keeping with the mechanical philosophy.

The notion of primitive force does double duty: it is the active power whose modifications are all the derivative active forces that result in the motions of bodies as they act on one another, but it is also the primitive passive power responsible for the resistance to such changes, namely resistance to penetration (antitypy), and the resistance to change of motion (inertia). The latter are manifested in bodies as derivative passive forces. The derivative active forces, on the other hand, are of two kinds: live forces[83] and dead forces. The live forces are manifested in the actual motions of bodies, and the dead forces in what may justly be called virtual motions. Thus, the forces at work in the balancing of weights or in elasticity are dead forces, the result of an infinite aggregation (integral) of infinitesimal tendencies to move, as are those producing gravity and centrifugal force. Live forces, on the other hand are the result of an integral of the product of velocity and mass through time, and if equal elements of velocity (i.e. endeavours) are produced in equal times (as they are in the case of falling bodies), the force produced is proportional to $\int mv\, dv$, and thus to mv^2. Both live and dead force, however, are conceived as instantaneous: their values signify how much force there is in the body at a given time. In the case of an ideal pendulum, for example, there is as much active force in the bob when it is at the top of its swing as there is in it at the bottom, even though it is not noticeably moving at that time. The live force mv^2 has been converted into an equivalent dead force, which is stored somehow in the bob when it is at rest at the top of its swing (we would say it is stored as the potential energy of the bob raised up in the gravitational field). In the phenomena of collisions, live force is gradually converted into an

[83] In translating *vis viva* as 'live force' rather than the more entrenched translation 'living force', we are following the lead of Jennifer Coopersmith (2017, 16). See the note on this translation in text [22], *A Specimen of Dynamics*.

equivalent dead force stored in the body through elastic deformation, and then reconverted back into live force as the body rebounds, consistently with the Law of Continuity. Leibniz explains this conceptualization of collisions in the unpublished second part of the *Specimen dynamicum*.[84]

One thing he declines to explain in the *Specimen*, however—although he does allude to it—is the a priori argument he has developed for the conservation of force on the basis of his new concept of *action*.[85] Instead, he repeats the a posteriori argument he has already refined in his public debate with Papin. According to that argument, the force expended in a violent motion[86] must be estimated by finding an effect that is homogeneous, that is, that can be divided into equal elements that are not simply modal, like velocity, but substantial. Thus 'the raising of a heavy body to two or three feet is precisely two or three times the raising of the same body to one foot' (neglecting the tiny differences in the force of gravity at different heights). Another acceptable effect might be the degree to which an elastic body is deformed, although this is harder to estimate. But simply doubling the velocity of the same body will not result in a doubling of its force, as measured by its capacity to produce a repeatable, homogeneous effect, such as the raising of a body of a given weight through a given height. Here Leibniz repeats the argument of the *Brief Demonstration*, as well as the reductio arguments against the Cartesian measure he had presented to Papin. These are based on the Principle of Equipollence, that the entire effect is always equal to the full cause; so if a pendulum bob falling through a height h, and colliding with an equal bob at the bottom of its swing gives all its force to it, where the force is taken to be equal to mv, then it will only have sufficient force to climb to a height of $\sqrt{h}$, according to Galileo's law; or, if the second bob is half the mass of the falling one, then when it receives all the force from the first bob it will have twice the velocity, according to the Cartesian measure, and therefore will be able to rise to four times the height, according to Galileo's law, thus enabling perpetual motion.

That argument, however, pertains to violent motion, where all the force is assumed to be expended in producing the effect. The a priori argument pertains to non-violent or unconstrained motion, for instance a motion with a constant velocity along a horizontal plane, where the body conserves its

[84] This second part of the *Specimen dynamicum* may be found in (GM VI 234–46/L 444–50).

[85] For discussion, see Duchesneau (1994, 183), Lodge (Leibniz 2013), and Garber and Tho (2018).

[86] It should be noted that in the seventeenth century the meaning of 'violent motion' had diverged from its original Aristotelian or Scholastic meaning of a motion that, unlike a 'natural motion', was not in a body's nature. Just as its counterpart 'natural motion' had come to be understood as a uniform motion in a straight line, so a violent motion had come to be understood as a non-inertial mechanical motion.

motion, so that its active force at each instant may be conceived as constantly producing what he calls an 'innocuous effect', which is proportional to the force and the time through which it acts. Now, the quantity of formal action produced by such a motion, he argues, depends on the mass ('quantity of matter', *m*) and the length *s* through which it is moved.[87] And, since the action expended in producing such an effect in half the time is twice as great, this means that the action is inversely proportional to the time, and thus proportional to the velocity *v*. So, according to this a priori argument, action is given by mvs, and thus by mv^2t. But since action is defined as force times time, this gives the measure of force as mv^2. It is not known why Leibniz did not include this argument in the *Specimen*, but he had difficulty persuading his correspondents to accept it.[88]

In the 1690s Leibniz was kept busy by the efforts of the Bernoulli brothers and also the Marquis de l'Hôpital to apply his calculus to problems in 'mixed mathematics', as well as to defend it against criticisms, such as those of Bernard Nieuwentijt,[89] and the imputations of Fatio de Duillier. He was involved in the solving and dissemination of multiple challenge problems, including (i) Vincenzo Viviani's so-called 'Florentine Problem' of 1692,[90] (ii) a tangent problem devised by Johann Bernoulli in 1693, (iii) the drawbridge problem of Joseph Sauveur,[91] (iv) Jacob Bernoulli's inverse tangent problem of

[87] 'The quantity of formal action in motion is that whose measure is instituted by the fact that a certain quantity of matter is moved along a certain length (the motion being uniformly equi-distributed) within a certain time' (*Dynamica* GM VI 345–6).

[88] Thus Jacob Bernoulli could not accept that an action produced in half the time is twice as great, insisting that since the effect is the same, so should be the action, and De Volder had similar qualms. See Paul Lodge's introduction to *The Leibniz-De Volder Correspondence* (Leibniz 2013) for discussion of Leibniz's attempts, with Johann Bernoulli, to persuade De Volder, and the main text of that volume for the sources.

[89] Bernard Nieuwentijt (1654–1718) was a Dutch mathematician noted for his objections to the Leibnizian calculus, and his attempt to found it instead on nil-square infinitesimals (infinitesimals whose squares and higher powers are all zero). This he did in Nieuwentijt (1694) and (1695). Unlike many other critics of Leibniz's calculus, however, such as Fatio de Duillier and Newton's acolytes, he did this in a respectful way, and Leibniz replied in kind in (Leibniz 1695).

[90] Vincenzo Viviani (1622–1703) was a mathematician and student of Galileo's. Leibniz had met Viviani in Florence in late November 1689, and had tried to assist him in getting the censure of Copernicanism withdrawn. See Bertoloni Meli (1988), and Arthur (2021, ch. 3 and Appendix 4) for discussion and translations of relevant texts by Leibniz. The so-called 'Florentine Problem' (see Roero 1990) was that of 'determining four equal windows on the surface of a hemispherical temple in such a way that the remaining part of the surface, after eliminating the windows, was equivalent in area to a square' (Roero 1990, 426).

[91] Joseph Sauveur (1653–1716) was a French mathematician and music theorist celebrated for his detailed work on musical nomenclature, divisions of the octave, and many founding concepts for the science of acoustics. The drawbridge problem was that of determining the curve that would be traced by a moving counterweight suspended by a rope over a pulley attached to one end of the drawbridge, such that the system remains in equilibrium for every position of the weight on the curve. It was first made public by Guillaume-François de l'Hôpital who provided a solution in his (1695) article in the *Acta Eruditorum*, where he attributed it to 'a certain geometer very expert in Cartesian analysis', now known to be Sauveur.

October 1694, and (v) the brachistochrone problem devised by Johann Bernoulli in 1696. All of this required not only the promotion of his methods, but responses to criticisms as well. Thus, although Jacob Bernoulli was one of the pre-eminent practitioners of the calculus (and had even coined the term 'integral calculus', also adopted by his brother Johann), he was constantly goading Leibniz with prompts, criticisms, and hints that Leibniz was perhaps trying to take credit for discoveries that were not his own innovations. An example of this was provided by circumstances surrounding the solution that Bernoulli had obtained to a challenge problem Leibniz had proposed earlier, which he had achieved as a happy by-product of his original research on the problem of the form taken by an elastic sheet or sail under the action of forces applied to its extremities. This had led him to his famous *elastic curve*, whose differential equation is $dy=x^2dx/\surd(a^4-x^4)$, and whose quadrature he had given as an infinite series (see notes on text [**23**]). But in working on this Bernoulli had noticed that the rectification of this curve gave the construction for the problem Leibniz had set in 1689 (in [**12**]) and which had since been forgotten, that of determining the curve traced by a body equably approaching or receding from a given point under its own weight, the so-called paracentric isochronous curve. Bernoulli gave his solutions to both problems in the June 1694 issue of the *Acta eruditorum*, but not without directing the latter to Leibniz's attention, inviting him, a touch ironically, to compare his solution with Bernoulli's. This provoked Leibniz to publish in the *Acta eruditorum* of August 1694 a major article[92] on the paracentric isochronous curve, in which, among other things, he made clear for the first time the need for an arbitrary constant in (what we call) an indefinite integration. Bernoulli then insinuated in a paper of December 1695 that Leibniz, by referring (in his 1694 paper) to his theorems on osculating circles, was trying to take credit for the discovery of the forms of the elastic sheet, and that he had only dealt with these problems after seeing Bernoulli's own solutions. This and other claims made by Bernoulli in his publications goaded Leibniz into dashing out a reply on a scrap of paper at the Brunswick Book Fair (according to his account), immediately on receiving his copy of the December *Acta eruditorum* there. The resulting 'Short Note' provides a concise encapsulation of some of these problems in mixed mathematics, and is given here as [**23**] in our collection.

Leibniz's calculus proved to be a great success in solving these problems in mixed mathematics, and greatly contributed to his prestige on the

[92] '*Constructio propria Problematis de Curva Isochrona Paracentrica*...' (Leibniz 1694).

European continent. To his obvious regret, however, his mathematico-physical contribution to the theory of celestial motions in the *Tentamen* did not attract any supporters except for the Bernoulli brothers, and his differential equation was ignored. Partly this was because Leibniz had failed to solve the two outstanding problems he had noted at the end of the *Tentamen*, namely to give a satisfactory mechanical explanation of gravitation, and to give a derivation of Kepler's Third Law. Leibniz had adopted the hypothesis of a vortex to explain how the celestial bodies were carried in the same plane by the aethereal fluid, and his harmonic vortex equipped them with a motion sweeping out equal areas in equal times in accordance with Kepler's area law. But the area law was assumed a posteriori, whereas Newton had neatly derived the area law assuming only a composition of inertial motion ('trajection') with a continuous gravitational attraction, whatever the law of attraction. On Leibniz's analysis, the 'paracentric motion' along the radius from the Sun at one focus of the ellipse was the result of the contest between the centrifugal endeavour resulting from the harmonic motion and the 'attraction' of gravity. To explain the latter he had tentatively offered his 'explosion' theory in 'On the Cause of Gravity', but this had no takers.

There were further problems, though, as his critics, notably the Oxford professor of astronomy David Gregory (1659–1708), were not slow to point out. One was that it left the orbits of comets not only unexplained, but with a motion inconsistent with the harmonic circulation.[93] Secondly, if the planets move according to a harmonic circulation where their velocities are as the inverse squares of their radii, their periodic times should be as r^2, contradicting the $r^{3/2}$ ratio of Kepler's Third Law, so that 'the different planets do not move by a harmonic circulation' (Gregory 1702, 102).[94] Another problem was pointed out by Leibniz's main ally in France, Pierre Varignon, who had been defending Leibniz's calculus against the attacks of Rolle and La Hire in

[93] See Gregory (1702, 101). Leibniz did not address this objection in his published reply, although in his unpublished essay, *Illustratio tentaminis*, he had replied to Gregory's objections point by point; there he alluded to the examples of twigs in an eddy in water taking different paths when thrown in together, and of different sound waves in the same place not interfering with one another (GM VI 266; Bertoloni Meli 1993, 187, n.29). Johann Bernoulli later gave an explanation of cometary orbits by suggesting further hypothetical vortices carrying the comets; see Eric Aiton (1960, 68).

[94] This criticism is in fact misplaced, since in harmonic circulation it is not the orbital velocities that are 'reciprocally proportional to the radii', as Gregory claims (1702, 102), but their components perpendicular to the radius. As Bertoloni Meli observes, 'the harmonic circulation imposes no constraints on the dependence of force with distance' (1993, 187); it is equivalent to Kepler's Area Law, so that any incompatibility with the Third Law would affect Newton's theory equally. The same misplaced criticism was later also made by John Keill (1671–1721), a Scottish mathematician who was part of Newton's circle, and a major contributor to the later bad blood between Newton and Leibniz.

the Parisian *Académie des Sciences*.[95] Pursuant on his publications on motions under a central force (1700), (1701), Varignon had pointed out in correspondence that in the *Tentamen* Leibniz had incorrectly represented the centrifugal force, thereby (he claimed) underestimating it by a factor of 2.[96]

Responding to these criticisms, Leibniz wrote an essay almost as long as the *Tentamen* itself, the *Illustratio tentaminis de motuum coelestium causis* (GM 254–76), which he submitted for publication in the *Acta eruditorum* in the autumn of 1705. The editor Otto Mencke rejected it as excessively long, however, so Leibniz had to content himself with publishing a short summary under the guise of an excerpt from a letter to Varignon.[97] This is [**24**] in our collection. In it he replied to some of Gregory's and Varignon's criticisms, correcting the error pointed out by Varignon, and at the same time other errors and misprints in the 1689 essay.

Our last selections concern Leibniz's controversy with Nicolas Hartsoeker (1656–1715). The dispute was initiated by the Dutch natural philosopher's *Conjectures physiques* (1706), a copy of which he sent to Leibniz with a request for comments. Leibniz obliged him with comments in a series of letters,[98] which resulted in Hartsoeker's publishing his *Eclaircissements sur les Conjectures physiques* in 1710. There he responds to some of these criticisms,

[95] Pierre Varignon (1654–1722) was a professor of mathematics in Paris who was elected to the Académie Royale des Sciences in 1688. He was the earliest advocate of Leibniz's calculus in France, and is noted for his presentation of the inertial mechanics of Newton's *Principia* using Leibniz's differential calculus, and for various contributions to mechanics and mathematics. Michel Rolle (1652–1719) was a mathematician known for his contributions to Diophantine analysis. A *pensionnaire géometre* in the Académie Royale, Rolle attacked the differential calculus for being inaccurate and based on unsound reasoning. Philippe de La Hire (1640–1718) was a third member of the Académie Royale, a painter as well as a natural philosopher and mathematician, and an early proponent of perspective and the beginnings of projective geometry.

[96] As Bertoloni Meli has explained (1993, 80–4), this is a complicated affair. Representing forces by the deviation from the tangent, in the case of gravity Leibniz had used a polygonal representation, taking the 'tangent' to be the prolongation of the chord. But in calculating centrifugal endeavour he had used the deviation from the actual tangent (that is, the 'versed sine') when it should have been twice this in the polygonal representation, namely the deviation from the prolonged chord. This results in his claiming that gravity is balanced by 'twice the centrifugal endeavour', instead of simply by the centrifugal endeavour. Varignon, for his part, had assumed that during an element of time there is a continuously acting force, as in Newton's Proposition 6 of Book 1, where Galileo's t^2 law is assumed to hold within each moment. In such a continuous representation this introduces a factor of ½ in the deviation of the orbital motion from the tangent, yielding (with some added geometry) the correct measure v^2/r for the centrifugal force. Leibniz, however, consistently rejected the idea that there could be an acceleration during a moment, as in the Newtonian representation. So, he persisted with the polygonal representation, where the deviation representing centrifugal force is given by the prolonged chord. Accordingly, he suggested replacing 'twice the centrifugal endeavour' by simply 'centrifugal endeavour' everywhere in the *Tentamen*, giving the same result, but with a corrected justification. See Aiton (1960) for an analysis of the exchanges between Leibniz and Varignon on this point.

[97] See the discussion of this episode in Antognazza (2009, 473–4).

[98] Leibniz's letters of 4 October 1706 (GP III 490–2), 10 March 1707 (GP III 492–4), 9 April 1707 (GP III 494–6) and 26 June 1709 (GP III 496) deal mainly with chymical and optical phenomena. Leibniz only briefly mentions his aversion to atoms in the letter of 9 April (GP III 495).

but without mentioning Leibniz by name. In his subsequent reply of 10 June 1710, Leibniz criticizes Hartsoeker's adoption of atoms, and this precipitates a more polemical exchange between them, centred on this subject.[99] Leibniz's friend, Bartholomew Des Bosses,[100] had acted as a conduit for their exchange, and considering it to be of general interest, urged Leibniz to publish selections from it. With Hartsoeker's consent secured, Leibniz's letter of 10 February 1711, and Hartsoeker's rejoinder of 13 March 1711, were published together in the *Mémoires de Trévoux* of March 1712 with introductory preface and postface by the editors (and then republished in December of the same year in the *Journal des Sçavans*). This article from the *Mémoires* constitutes item [**25**] in our selection. Subsequently Leibniz's reply to Hartsoeker of July 1711 was also published in the *Mémoires*—the 26th and last item in our selection [**26**]—but no further part of their correspondence.

This controversy is of interest for showing how Leibniz maintained many views that he had set out early in his career: his rejection of atoms (already in place in 1676), his explanation of cohesion in terms of 'conspiring motions' (dating from his *Hypothesis physica nova* of 1671), and his thesis that perfectly hard atoms would be incapable of reaction or perception (in place by 1679–80). Hartsoeker's views on a vital first element (reminiscent of the earlier atomism of Sébastien Basson) also provide an interesting contrast to Leibniz's philosophy of monads or 'living atoms', and indeed to the views on atoms that Leibniz himself had entertained up to 1676.[101]

The correspondence also features an implicit attack on Newton's universal gravitation, through the analogy Leibniz draws between Hartsoeker's hypothesis of perfectly hard atoms and Newton's hypothesis of a universal attraction between all massive bodies: in neither case, Leibniz charged, is a naturalistic explanation provided for these alleged phenomena. This criticism was brought to Newton's attention by John Keill when an English translation of the first of these articles from the *Mémoires de Trévoux* appeared in the *Memoirs of*

[99] Of this ensuing correspondence, Gerhardt reproduces letters from Hartsoeker to Leibniz of 8 July, 22 August and 30 December 1710, 13 March 1711, and 6 June 1712, and from Leibniz an undated extract of a letter of July/August 1710*, and letters of 30 October 1710, 10 February, 9 July, and 7 December 1711*, and 8 February 1712 (GP III 498–535). English translations of these * pieces by Strickland available at http://www.leibniz-translations.com/hartsoeker1710.htm and http://www.leibniz-translations.com/hartsoeker1711.htm.

[100] Bartholomew Des Bosses (1668–1738) was a Jesuit priest who befriended Leibniz in 1706, offering to help reconcile Leibniz's views with the Aristotelian views espoused by his order. He also made a Latin translation of Leibniz's *Essais de Théodicée*. See Look and Rutherford's introduction to (Leibniz 2007).

[101] For an analysis of the development of Leibniz's views on atoms and souls in relation to these earlier traditions, see Arthur (2018, chs 3 and 4).

Literature of London, 5 May 1712.[102] Newton was incensed, and drafted a stinging reply to Leibniz's criticisms of the idea of universal gravitation, although this was not published during his lifetime. Instead, the task of defending the Newtonian philosophy against Leibniz's criticisms was taken up by his friend Samuel Clarke, issuing in the influential Leibniz–Clarke correspondence, published a year after Leibniz's untimely demise.[103]

In conclusion, it is hoped that this selection of Leibniz's published articles gives a fair representation of his efforts and accomplishments in natural philosophy, and how they were received in his time. As such, it should help to redress the balance in favour of his published work, despite the treasures that are still being unearthed from the archives. As we have witnessed, Leibniz was keen on sharing his views with his contemporaries in an effort to advance knowledge. Indeed, one could say that he regarded cooperative work of this kind as more important than the publishing of monographs, thus giving added significance to selections of his published articles such as this one. The fact that he published over a hundred journal articles is a testament to this desire for the promotion of knowledge, and should be borne in mind whenever it is charged that Leibniz 'published so little in his lifetime', as is often said.

[102] As noted in text [**25**], Keill had already stirred up trouble by accusing Leibniz in print of plagiarizing the calculus from Newton. So when the Royal Society was supposed to be adjudicating the dispute, he made sure that Newton was apprised of this criticism.

[103] The most complete edition of the Leibniz–Clarke correspondence in English remains that of H. G. Alexander (Leibniz 1956), which includes Clarke's translation of Leibniz's French letters into English, all his replies, and the extracts he made from the published writings of Newton and Leibniz.

1

A Unitary Principle of Optics, Catoptrics, and Dioptrics[1]

G. W. Leibniz, *Acta eruditorum*, June 1682[2]

The *primary hypothesis* common to these sciences—by means of which the direction of any ray of light is determined geometrically—can be put in this way: *Light from a radiating point reaches an illuminated point by the easiest path, which is to be determined first with respect to planar surfaces, but is accommodated to concave and convex surfaces by considering their tangent planes.* Still, I do not have here an account of certain irregularities that perhaps play a role in the generation of colours and in other extraordinary phenomena, which are not attended to in the consideration of optics.

Hence *in simple optics, the direct ray from the radiating point C reaches the illuminated point E* by the shortest direct path—in the same underlying medium of course—that is *by the straight line CE* [Fig. 1.1].

In Catoptrics the angle of incidence CEA and the angle of reflection DEB are equal.[3] For example, let *C* be the radiating point, *D* the illuminated point, and *AB* a plane mirror: the point *E* on the mirror is sought at which the ray is reflected to *D*. I say that it is such that as a result the whole path *CE* + *ED* becomes the least of all, or less than *CF* + *FD* if, that is, a different point *F* on the mirror had been assumed. This will be obtained if *E* is taken to be such that

[1] From the Latin: *Unicum Opticae, Catoptricae & Dioptricae Principium, Acta eruditorum*, June 1682, 185–90. (ESP 37–43.) Translated by Jeffrey McDonough, Lea Schroeder, and Samuel Levey.

[2] This paper is the second in a series of papers Leibniz submitted to the newly established journal at the request of one of the editors, Christoph Pfautz. In it he follows up his *De vera proportione circuli* of February (a paper in pure mathematics) with one in applied mathematics, featuring an application of his 'method of maxima and minima' to optics. He situates the argument as depending on Fermat's Principle of Minimal Path, although Leibniz interprets that principle in terms of the path offering least difficulty, rather than the path taking the least time. Here, however, he only reports the result of the calculation, and does not reveal how he used his new calculus to obtain it until his publication of his *Nova Methodus* paper in the *Acta eruditorum* of 1684.

[3] Today we designate the angles of incidence and reflection not with respect to the horizontal surface, as Leibniz does here, but with respect to the perpendicular to the surface. 'Catoptrics' is that branch of optics that deals with reflection; 'dioptrics' that branch concerning refraction.

Leibniz: Journal Articles on Natural Philosophy. Richard T. W. Arthur, Oxford University Press.
DOI: 10.1093/oso/9780192843531.003.0002

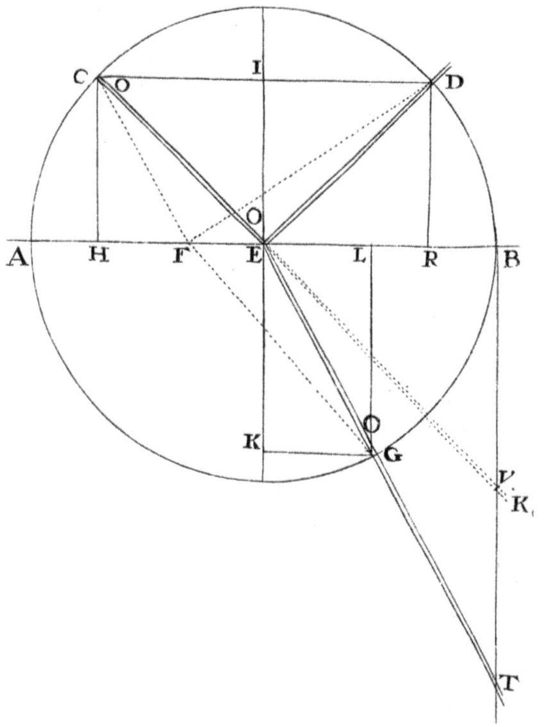

Fig. 1.1

the angles *CEA* and *DEB* are equal, as is evident from geometry. Ptolemy[4] and other ancients used this demonstration and it is found in Heliodorus of Larissa[5] and elsewhere.

In Dioptrics, the complementing sines, EH and EL,[6] *of the angle of incidence CEA and the angle of refraction GEB, always preserve the same ratio, which is reciprocal to the resistance of the media.* Let *IE* be air, *EK* water, or glass, or some other medium denser than air, *C* a radiating point in the air, *G* an illuminated point beneath the water: it is asked by which path the light shines from the former to the latter; ôr, what is the point *E* on the surface of the water *AB* at which the ray emitted from *C* is refracted and sent to *G*? This point *E*

[4] Here is meant the *Catoptrics* of Hero of Alexandria (*Opera*, II, 1, 1900, pp. 316–65), a work that was for a long time wrongly ascribed to Ptolemy.

[5] This is the *Capita opticorum*, Cap. 14, of Damianos of Larissa, attributed to his father Heliodoros on the basis of the title found in some manuscripts.

[6] In the seventeenth century sines were interpreted as lines: in a circle of radius 1, the sine of the angle between two radii would be the perpendicular dropped from the intersection of one of them with the circumference onto the other; in other words, the sine can be understood as the ratio of the opposite to the hypotenuse with the latter tacitly taken as 1.

should be taken in such a way that the path is the easiest of all. Now, in different media, the difficulties of the path are in the compound ratio of the length of the paths and the resistance of the media. If the straight lines m and n represent resistance with respect to light—the former of air, the latter of water—the difficulty of the path from C to E will be as the rectangle formed by CE and m, and the difficulty of that from E to G will be as the rectangle formed by EG and n.[7] Therefore, in order that the difficulty of the path CEG be the least of all, [185/186] the sum of the rectangles CE times m + EG times n should be the least possible, or less than CF times m + FG times n for any point F other than E.

We are trying to find E. Since therefore the points C and G and also the straight line AB are given by supposition, the straight lines perpendicular to the plane—we will call CH, c; and GL, g; and HL, h—are therefore given as well. And the line EH that was sought we will call y, EL will be $h - y$, and CE will be $\sqrt{(c^2 + y^2)}$, which we will call p, and EH will be $\sqrt{(g^2 + y^2 - 2hy + h^2)}$, which we will call '$q$'. Therefore, $m \sqrt{(c^2 + y^2)} + n \sqrt{(g^2 + y^2 - 2hy + h^2)}$, that is $mp + nq$, should be the least of all possible quantities similarly expressed, and y is sought accordingly. *By means of my method of maxima and minima*, which above all notations thus far marvellously shortens the calculation, it is clear right away at first glance—almost without any calculation[8]—that mq times y will be equal to np times $h - y$, ôr that np will be to mq as y will be to $h - y$, ôr that the rectangle CE by n will be to the rectangle EH by m, as EH will be to EL. Therefore, assuming that CE and EG are equal, n (the resistance of the water with respect to light) will be to m (the resistance of the air) as EH (the sine of the complementing angle of incidence CEA in the air) is to EL (the sine of the complementing angle of refraction GEB in water), ôr the complementing sines will be in a reciprocal ratio to the resistance of the media—as was claimed. And so if EL is discovered to be $^2/_3$ of EH in one example or experiment, it will also be so in all other cases in which C and G are taken to be in air and in glass, respectively. If E is in air and G underwater, EL will be approximately $^3/_4$ of EH.

We have therefore reduced all the laws of rays confirmed by experience to pure geometry and calculation by applying one principle, taken from final

[7] This is a typical example of the geometric expression of algebraic facts common in the seventeenth century, as discussed above in the section 'Some Notes on Seventeenth-century Mathematics'.

[8] Leibniz refrains from actually giving the calculation using differentiation here, which he will only do in his 1684 paper. In order to minimize $mp + nq$, he is setting $d(mp + nq) = 0$. Since $p = \sqrt{(g^2 + y^2 - 2hy + h^2)}$, $dp = 2y/2\sqrt{(g^2 + y^2 - 2hy + h^2)}dy = y\ dy\ /p$. Similarly, with $q = \sqrt{(g^2 + y^2 - 2hy + h^2)}$, $dq = -\ (h - y)\ dy\ /q$, giving $mqy = np(h - y)$.

causes, if you consider the matter correctly: for a ray setting out from *C* neither takes into account how it could most easily reach point *E* or *D* or *G*, nor is it directed to these through itself, but the Creator of things created light in such a way that this most beautiful event would arise from its nature. And so those who with Descartes reject *final causes* in physics[9] err greatly—not to say something more offensive—since, besides the admiration of divine wisdom, these causes provide us with the most beautiful *principle* for also *discovering* the properties of those things whose interior nature is still not so clearly known to us that we are able to use proximate efficient causes and explain the machines which the Creator employed for producing those effects and for obtaining his ends. Hence we also understand that the meditations of the ancients on these matters as well are not to be so looked down upon as it seems some people do today. [186/187] For it seems to me very likely that the great geometers Snell and Fermat—very well versed in the geometry of the ancients—extended the method that the ancients had used in Catoptrics to Dioptrics. Indeed, I suspect that Snell's theorem—which that most distinguished man Isaac Vossius cites from Snell's three unpublished books on optics[10]—was discovered by almost the same method (although I believe not by such a felicitous calculation as we have used here). Indeed, from our method it follows immediately, as is to be demonstrated. Let a circle *CBG* be described with the centre *E* and radius *EC* or *EG*; the extended tangent to this [circle] at *B* meets *CE* at *V* and *EG* at *T*.

If an eye is at *C* and an object that is seen beneath the water is at *T*, the point *T* will appear to be at *V* because it seems to us that we see along the straight line *CEV*, but since we in fact see by the broken line *CET*, it is clear that *EV* is the secant of the angle of incidence *CEA* or of *VEB*, which is equal to it, and that *ET* is the secant of the angle of refraction *GEB*. From a known proposition of trigonometry, moreover, secants are reciprocal to their complementing sines, and therefore it follows directly that *EV* is to *ET* as *EL* is to *EH*, ôr (by our theorem) as *m* is to *n*. Therefore, given that the eye at *C* is present in a different medium than the object at *T*, the apparent ray *EV* in the medium of

[9] Cf. Descartes, *Principia Philosophiae*, III, 2: 'We must beware of being so presumptuous as to think we understand the ends which God set before himself in creating the world' (AT VIIIA, 81). These remarks are directed against Cartesians like Clerselier, who claimed that appealing to final causes was tantamount to ascribing knowledge and choice to nature. See Clerselier to Fermat, May 1662 (F II 462), and Sabra's discussion (Sabra 1981, 153).

[10] Isaak Vossius (or Isaac Voss) (1618–1689) was a Dutch scholar and critic of Descartes, highly reputed in his own time, who published his *De lucis natura et proprietate* (1662). The main claim of the book today is that in its chapter XVI Vossius brought to the attention of the scholarly community that Willebrord Snell had discovered the law of refraction 15 years before Descartes published it in his *La dioptrique* in 1637.

the object (water) will be to the true ray *ET* in the medium of the object (water) as the resistance *m* of the medium of the eye (air) to the resistance *n* of the medium of the object (water). This ratio is always the same provided that the media are the same, and therefore the ratio between the true ray *ET* and the apparent [ray] *EV* will always be the same, which was Snell's theorem.

In the same way, the ratio of the sines of the complements of the angles of refraction and incidence, *EL* and *EH*, will always be the same, since for us it is reciprocal to the resistance of the media: that is the theorem of the Cartesians, although Descartes understood the resistances of the media differently—indeed, inversely—from us. Therefore, upon noticing this agreement of conclusions, the most distinguished Spleiss, who is also very well versed in these studies, wondered (not without ground)[11] whether Descartes might not have seen Snell's Theorem when he was in Holland; indeed, Spleiss notes that Descartes himself had regularly omitted the names of authors and gives the example of world-vortices which Giordano Bruno and Johannes Kepler[12] came within a finger's breadth of attaining, so that all that seemed to be missing was the name. He adds that Descartes falls into great difficulties because he wanted to demonstrate this theorem using his own tools: indeed, he suspected that the ray *CE* meets with less resistance in water or glass than in air, since he saw that upon entering the water from the air it is refracted into *EG* and so towards the perpendicular *EK*, and is therefore rendered more similar to that ray whose action is stronger, namely, the perpendicular. Nonetheless, in supposing the contrary [187/188] which is much more consistent with reason, the same conclusion is reached upon applying our principle of the easiest path. Fermat rightly concluded from this that Descartes had not given the true reason of his own theorem. Furthermore, the analogy by which the latter tries to illustrate his own explanation is not very apt. In *fig. 2* [Fig. 1.2] let there be a little globe *A* on a polished table *BC* advancing in the place $_1A$: let this, in the middle of its course, run into the part of the table *DE* covered with a woollen cloth, whereby it will move more slowly in $_2A$.

Consequently, he thinks that in the same way glass or another solid body will delay rays of light less than air, which is more fibrous [*villosus*]. But (passing over [the fact that] even parts of water are *sufficiently pliable*

[11] Hess and Babin (Leibniz 2011, 23, n. 10) note that, in a letter to Gerhard Meier of 26 October 1690 (A 1, 6 N. 127), Leibniz remarks that Spleiss had already mentioned this before June 1682. This is an implicit reference to Stephan Spleiss's *Observatio CXCI* in the *Miscellanea curiosa* of 1679, which bears the title *De visionis distinctissimae loco*.

[12] See, respectively, the end of the second dialogue of Bruno's *De l'infinito universo et mondi*, 1584, and Kepler's *Astronomia nova*, 1609, chapter XXXIV.

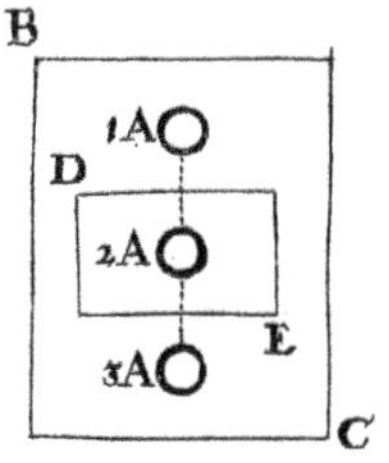

Fig. 1.2

according to Descartes himself) it is sufficient to consider that when the little globe—having moved from $_2A$ beyond the cloth *DE*—reaches once more a polished part of the table in $_3A$, it does not at that point recover the prior speed that it had at $_1A$ before it encountered the cloth. By contrast, when a ray of light [travelling] from a medium of greater resistance into a medium of less resistance has entered again into a medium similar to the earlier medium, it would recover its earlier state. Supposing that the first and last surfaces of the two similar media (the former being that of the emitting medium, and the latter that of the receiving medium) are parallel planes, then as a result of the later refraction the light ray would recover a direction parallel to the one it had before the first refraction.

Nevertheless, it seems that how Descartes explains—in a way worthy of his genius—the reflection as well as the refraction of light by an imitation of the motion of other bodies, is not to be rejected, but only emended. With regard to reflection, he should have first explained why some globe such as *I*, following along the perpendicular *IE*, and striking the plane *AB*, is thereby reflected. Indeed, we see that some bodies, for instance soft ones, are not equally reflected. The true cause of this reflection is the elasticity of the globule, or the plane, or both. For an elastic plane will yield to some degree, as we see happen when a small stone strikes a stretched membrane or an inflated bladder; and indeed, the harder it is struck the more it will yield, and in restoring itself with a correspondingly greater force, it throws back what had struck it along the same path and with the same speed with which it had come. Indeed, although Descartes did not want to accept this explanation of reflection already brought forward by certain people in his own day, as is clear from his letters, today it has been established beyond doubt by arguments and experiments. Since therefore the globe comes from *C* to *E* along the straight line *CE* in *fig. 1*, it also for that reason comes by the motion composed from two motions—a horizontal one such as *CI* or *HE*, by which it comes from *CH* to *IE*, and a perpendicular one such as *CH* or *IE*, by which it comes from *CI* to

HE [188/189]—both setting out from *C* and terminating at *E*; and since the surface *AB* is not opposed, but rather parallel, to the horizontal endeavour of this motion arriving at *E* from *CH* toward *IE* along the straight line *CI* or *HE*, it will for that reason retain the same undiminished speed and direction of the horizontal motion, and—assuming that the interval between *CH* and *IE* is equal to the interval between *EI* and *RD*—it will take as much time to come from *EI* to *RD* as it did to come from *CH* to *IE*.

However, if the speed of the perpendicular motion by which it comes from *CI* to *HE* is undiminished, it will be reflected in the opposite direction, so that it will take the same amount of time to return from *ER* to *ID*. Therefore, since *ER* is equal to *EH* and *RD* to *CH*, the triangles *CHE* and *DRE* will be similar and equal and hence the angles *DEB* and *CEA* will be equal. All this will be clearer if we imagine that the segment *CID* parallel to the surface *AB* touches *AB* along *HR*, having remained parallel through *CH*, *DR*; meanwhile the globe on the segment *CI* is carried from *C* to *I*, so that in fact the whole composite motion of the globe will be along the diagonal *CE*. But the segment *CID*, having been reflected from the solid surface *AB*, will return with the same speed and direction by which it had come, and in an equal amount of time it will arrive again at *CID*. Meanwhile, the globe has continued to move along the ruler with the same speed, and therefore travels from *I* to *D* in the same amount of time, [or] along *ID* which is equal to *CI*—for equal spaces are crossed in equal amounts of time if the same speed is held fixed. And so the globe is carried from *E* to *D* along the straight line *ED* by a motion composed of its own motion along the ruler through *ID* and the motion of the ruler itself along *EI*, that is, its return from *HR* to *CD*.

In explaining refraction, it should be noted that a medium more resistant to light (yet still not opaque) seems to be that which impedes the diffusion of light more, ôr its distribution through more parts of the medium, and one can say that it is less illuminable, for it is the nature of light to try to diffuse itself. Conversely, the more that light will affect equally the parts of the medium that it illuminates, or where it will communicate its own force to multiple insensible particles of the illuminated place, the more the medium will be illuminable and the less it will be resistant to light. Hence by as much as the affected parts of the illuminated medium are more solid and small, or less interspersed with some other heterogeneous material not affected by light, to that extent the medium will be said to be more illuminated. Indeed, it is known from principles of Mechanics that the same blow impressed at the same time on multiple bodies will impart less force to the individual ones than if it had been inflicted on one of them; therefore, it happens that the more resistant a medium is to the diffusion of light, [189/190] ôr the fewer parts are affected, the more strongly the single parts will be affected; in a more illuminable medium, more parts are affected, but less strongly, and the impressed impetus is weaker. Now assuming that the motion of the little globe takes place along a radius and that the little globe comes from *G* to *E*, and there enters the [new] medium, its speed or impetus would be retarded in the proportion of, say, 3/2. Therefore, if the ray comes from *G* to *E* in the first medium *KE* in one unit of time, the same ray in the new medium *EI* will come from *E* to *C* in 3/2 units of time, assuming that *EC* and *GE* are equal, wherever *C* ultimately is. But since the surface *AB* separating the media will not obstruct the horizontal speed exercised on *GK*, *LE*, and parallel lines before the first entry of the little globe into the new medium *EI* or at the point *E* (indeed, the horizontal motion *LEH* only grazes it), the inclination of the line *CE* must be determined at once in the first moment of entering [the medium], or at the point *E* (considering the little globe to be indefinitely small like a point, just as rays are customarily viewed as being without breadth): hence such an inclination is to be assumed at the beginning so that with respect to the horizontal motion the velocity remains the same, and in the new medium it remains the same as when it first entered. Therefore, the little globe, which while it was moving in the prior medium from *G* to *E*, had completed in one unit of time an interval *GK* or *LE* (between *GL*, *KE*) in the horizontal direction, now has one and a half units of time during which it must go from *E* to *C*, so that it will cover the interval *EH* or *IC* (between *EI* and *HC*) in the same horizontal direction which must be one and a half times the prior interval *LE* because in keeping the same horizontal velocity (which the moment of refraction does not change) the distances are as times. Therefore *EH* is to *EL* in direct proportion of the times, or in reciprocal proportion of the speeds, or the resistances. For we have shown that in the case of light, due to the resistance of the medium impeding diffusion, the velocity or impetus increases in proportion to the resistance, and languishes in proportion to the ease with which it diffuses itself through single particles. Conversely, the ray recovers its force and also direction when it returns again to the medium where it is diffused less, and where more rays are spent on driving fewer parts. Descartes was unable to explain that recovery—as we have mentioned above—by his own comparison with a woollen cloth or some other fibrous material.

2

New Demonstrations Concerning the Resistance of Solids[1]

G. W. Leibniz, *Acta eruditorum*, July 1684[2]

The science of mechanics appears to have two parts, one concerning the power of acting or moving, the other concerning the power of being acted upon or resisting, that is, the strength of bodies. The latter of those has been discussed by only a few. Archimedes, who nearly alone among the ancients applied Geometry in Mechanics, did not touch this part. After Archimedes almost nothing was accomplished in Mechanical Geometry until Galileo, who, equipped with precise judgement and great knowledge of the inner recesses of Geometry, first extended the limits of this science and began likewise to apply the laws of Geometry to the resistance of solids. And although he did not hit the nail on the head either in this matter or in the matter concerning the motion of projectiles, since he used insufficiently certain hypotheses, nonetheless he reasoned correctly from the foundations he laid down.

Therefore, he understands the resistances of beams fixed to city or house walls in this way. [319/320] In *figures* 1 and 2 [Figs 2.1, 2.2], let the beam *ABC* be fixed perpendicularly into a wall or little support *DE*. Let *AC* be equal to *AB*, and in *fig.* 1, let a weight *F* be suspended at *C* which could precisely wrest a horizontal beam directly from the erected wall; and in *fig.* 2, a weight *G* which could precisely tear the horizontal beam from the perpendicular wall; and in

[1] From the Latin: *Demonstrationes Novae de Resistentia Solidorum autore G. G. L., Acta eruditorum*, July 1684, 319–25. (ESP 55–58; GM VI 106–12). Translated by Jeffrey McDonough, Lea Schroeder, and Samuel Levey.

[2] This paper is the result of an epistolary collaboration with the experimental physicist Edme Mariotte between August 1682 and January 1683. Leibniz and Mariotte had already agreed that matter—including air—is inherently elastic, and on this basis Leibniz had already made great progress in acoustics in collaboration with Günther Christoph Schelhammer (see the Introduction). On the basis of their shared assumption of the elasticity of all bodies, and specifically Mariotte's construal of bodies as comprised of elastic fibres, Mariotte had corrected Galileo's estimate of the ratio between the transverse and direct fracture of materials from a half to a quarter, and Leibniz had in turn corrected Mariotte's estimate instead to one third, as described here. This theoretical demonstration was accepted by Mariotte, who then proceeded to confirm it experimentally, publishing the result in his *Traité du Mouvement des Eaux* (1686). See O'Hara (forthcoming, introduction).

Leibniz: Journal Articles on Natural Philosophy. Richard T. W. Arthur, Oxford University Press.
DOI: 10.1093/oso/9780192843531.003.0003

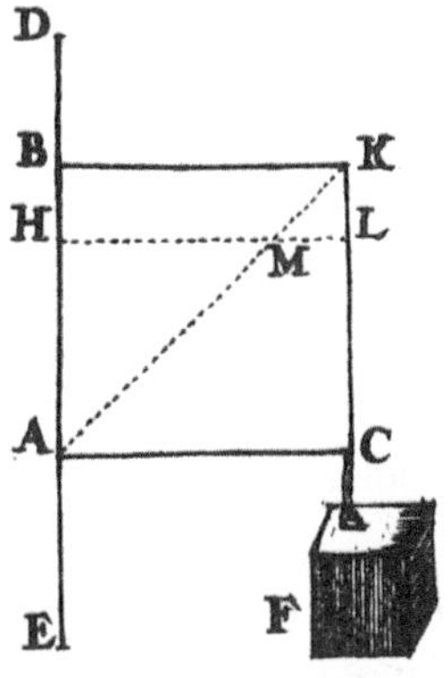

Fig. 2.1

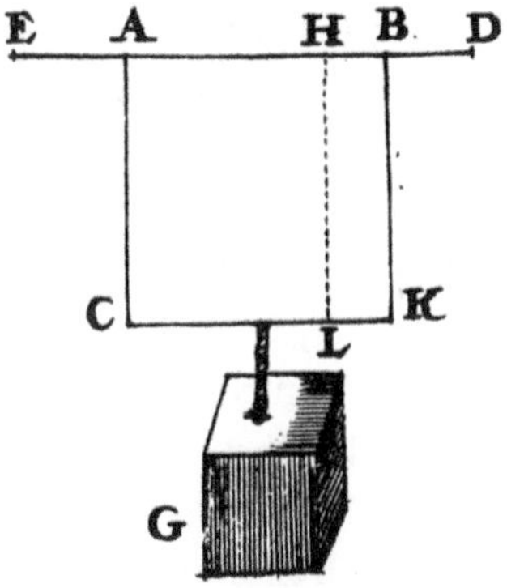

Fig. 2.2

fig. 2, a weight *G*, which could precisely tear the beam vertically from the horizontal support. (I will call the former case *transverse breaking*, the latter case *direct tearing*.) According to Galileo, the weight *F* will be half the weight of *G*, supposing the solid is perfectly rigid, that is, incapable of bending at all, and the weight of the beam itself is ignored or already included in the suspended weight. For since *AB* and *AC* are equal, it follows that the weight *F* in *fig.* 1 will meet with the same resistance at point *B*, as if it pulled perpendicularly, as in *fig.* 2. Therefore, the resistance at point *B* in either figure may be represented by *BK*, and so the resistance of point *H* in *fig.* 2, will be represented by *HL*, equal to *BK*, since in *fig.* 2, the resistance of all the points is the same. And the resistance of the same point *H* in *fig.* 1 will be represented by the ordinate *HM* applied to the triangle *ABK*, because—by the nature of a lever—it stands to the resistance of *B*, as *AH* to *AB*. And doing the same thing we did at point *H* at any other point whatsoever between *A* and *B*, the square *BC* will be completed in order to represent the resistance in *fig.* 2, and the triangle *ABK*—half the square—in order to represent the resistance in *fig.* 1. Thus, if the weight *F* is supposed to be precisely equal to the resistance in *fig.* 1, so that it would be overcome by adding any weight however small, it will also be half of *G* (precisely equal to the resistance in *fig.* 2). That is, the power of transverse breaking will be half the power of direct tearing (we will soon show that it is not really a half, but a third). From this already many practical conclusions can be deduced.

But these opinions of Galileo, and others of the same kind, were once examined by Paul Würz,[3] famous for his high military honours and actions some time ago, and who understood these studies very well. He attempted to test them by many experiments, but his results hardly corresponded to certain [of Galileo's] conclusions. I have it in just this way from the most distinguished Blondel,[4] excellent in these and other studies, recently Teacher of Mathematics to the Most Serene Dauphin, and director of the Academy of Architecture, who developed the same argument, and was familiar with Würz. But also the most distinguished Mariotte[5] from the Royal Academy, deservedly famous in optics and mechanics, discovered by performing experiments that a weight *F* much smaller than what Galileo wanted sufficed to break off the beam. The reason for this can be nothing other than [320/321] that Galileo considered a perfectly rigid beam, which would break off as a whole all at once when its resistance is overcome, even though all the bodies that have been given to us to treat yield to some degree before they can be torn off. Whence the most distinguished Mariotte, observing this, deduced by a clever calculation that the weight *F* is about a fourth part of the weight *G*. But once I had the opportunity to consider the matter more carefully, and to apply the laws of geometry to it, I seized on the true proportions and demonstrated, among other things, that the weight *F* will be one third the weight of *G*, and hence the strength of bodies resisting breakage is one and a half times less than the proportion Galileo wanted.

In order to understand this, one must recognize first of all that two cohering bodies are not immediately torn apart from one another in an instant; which

[3] Baron Paul von Würz (1612–1676) was a German soldier and diplomat who served in various capacities in the imperial, Swedish, Danish, and Dutch military and government services.

[4] François Blondel (1618–1686) was a famous French architect, military and civil engineer, and mathematician, as well as a soldier and diplomat. While serving in Italy he claimed to have studied with Galileo personally, and later became one of his earliest supporters in France. He was admitted to the *Académie des Sciences* as a geometer (cartographer) in 1669.

[5] For details about Mariotte and Leibniz's collaboration with him, see the Introduction.

can be judged by the example of a walking stick that bends before it breaks, and a string that stretches before it snaps; and the bending of the stick itself is a certain extension of its curved surface. And that there is nothing so rigid that it cannot be bent to some degree by even a light impact follows from the nature of sound, which is a certain tremor, or bending back and forth of the parts of the body emitting the sound, although the shorter and tenser the trembling parts, and the harder the body they constitute, the quicker and more insensible the restoration and the sharper the sound. The long and thin filaments of glass show that it is itself flexible; just as the Florentine experiments show that glass that is sufficiently thick is contracted by the cold. Indeed, we are taught by the senses that the parts of plants & animals are in a certain way woven, and consist of various intertwined filaments. Minerals and metals as well, although they were fluid, later congeal, and can now also have tenacity, be drawn into threads, be extended by a hammer, and adhere in fusion.

Therefore, let us consider the matter as if there are certain fibres which connect the parts of bodies, and let us understand the beam *BC* to be connected to the wall or support *DE* by many plaits of fibres bound at points *A*, *H*, *B*, and at countless other points in between. Once the weight *F* has been hung, the beam will be moved a certain amount around the fulcrum *A* in *fig.* 3 [Fig. 2.3]; and as the point *B* of the beam moves away from the point $_1B$ of the wall it will come to the point $_2B$ at a distance from the wall, and, drawing with it the fibre by which it is attached to the wall, it will stretch the fibre like a

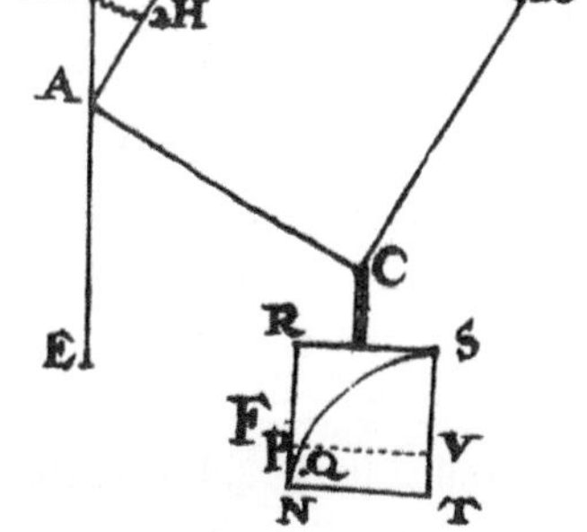

Fig. 2.3

chord, i.e. it will extend it beyond its natural state into the line ${}_1B_2B$; in the same way, point H will stretch its fibre into the line ${}_1H_2H$—lines which, although they are in fact imperceptible, are nevertheless represented visibly for the sake of instruction—[321/322] and indeed there will be less resistance to being drawn apart by the fibres at ${}_1H_2H$ than by the fibre ${}_1B_2B$, and that in double the ratio of the distance from A, that is, from the given distance taken twice.

For, *first*, the weight at C, needed to stretch the fibre ${}_1H_2H$ as much as ${}_1B_2B$ will be less than the weight required to stretch the fibre ${}_1B_2B$ by the ratio AH: AB. For example, if AH is a third of AB, then also the weight at C, which can stretch out the single fibre ${}_1H_2H$ so that it becomes equal to ${}_1B_2B$, will also be a third of the weight stretching the single fibre ${}_1B_2B$. But indeed, *second*, when both fibres are stretched together by the weight hung at C, the fibre ${}_1H_2H$ is not, of course, stretched as much as the fibre ${}_1B_2B$, but much less, and this again in the ratio AH:AB. For if AH is a third of AB, ${}_1H_2H$ will be a third of ${}_1B_2B$. And so (from the hypothesis confirmed elsewhere that extensions are proportional to stretching forces) the weight that will be needed to stretch ${}_1H_2H$ in this way will be only a third part of the weight that would have been needed to stretch it as much as ${}_1B_2B$. That is, a third of a third of the weight stretching ${}_1B_2B$, ôr a ninth.

Thus, in general, in this simultaneous tension of all the fibres appearing at any point, the resistances will be in the duplicate ratio of the distances taken from the lowest point of the fulcrum, ôr from the centre or axis of the balance; that is, the resistance at H will be to the resistance at B as the square of AH is to the square of AB. Thus, if we now let the weight F in *fig.* 3 be the parabolic body NRSQN, hung freely from C, in which the height NR is equal to the base RS (just as AB is equal to AC), and let the ordinates be proportional to the squares of the heights, i.e. if PQ is to RS as the square of NP is to the square of NR—then, supposing that the base RS represents the resistance at B, the ordinate PQ will represent the resistance at H, if, of course, the heights NP and NR are proportional to the corresponding heights AH and AB.

In fact, the whole three-sided concave parabolic figure $NRSQN$ will represent the resistance of the whole line AB, if, of course, the beam ABC is pushed down transversely, or in the manner of a lever, by a hung weight F. And the square $RNTS$ circumscribed around this three-sided parabolic figure would represent the direct resistance of the same line AB if, of course, the beam were to be torn directly from the wall as in *fig.* 2. For, since AB and AC are equal, the transverse resistance at point B will be the same as the direct resistance, represented, of course, by RS in *fig.* 3: now, if the beam is torn directly (as in

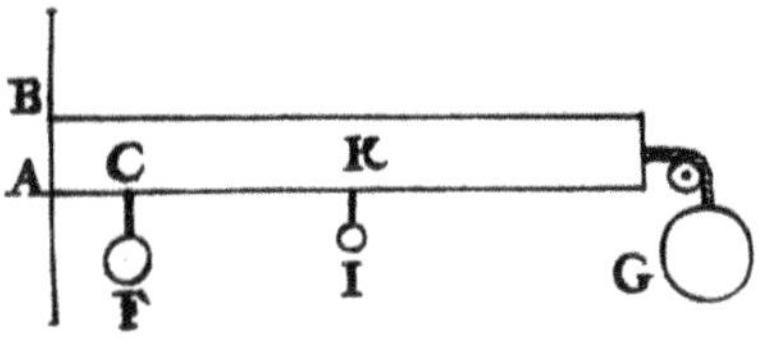

Fig. 2.4

fig. 2) the resistance at all the points is the same, then the direct resistance [322/323] at point *H* will be *PV*, equal to *RS*: and so by proceeding in this way through the other points the square *RT* will be completed, and it will then be three times the inscribed three-lined concave parabolic figure, namely, *NRSQN*, and therefore the direct resistance of any straight line (such as *AB*) will also be three times the transverse resistance.[6] QED.

Furthermore, whatever the length of the beam might be, or the distance of the hung weight from the wall (which thus far we have assumed to be equal to the height of the beam), it will be possible to easily determine the weight sufficient to break off the beam: if the weight *G* can directly tear the beam off as in *fig.* 4 [Fig. 2.4], then the weight *F* will be a third of *G* (provided that *AC* is equal to *AB*); but if the weight *I* is suspended from *K*, and *AK* is four times *AB* or *AC*, then the weight *I* will be a fourth of *F*, and a twelfth of *G*.

Generally, therefore a weight tearing off a parallelepiped beam will be to a weight transversely breaking <it> off, that is, in the manner of a lever, as the length of the lever is to a third of the thickness of the beam.

Thus far, however, we have considered the beam as itself lacking weight, but if its weight must enter into the account, then it will be as if a weight *I* equal to that of the beam were suspended from *K*, the centre of gravity of the beam.

It could even happen that a beam should break under its own weight in some place, such as *G* in *fig.* 5 [Fig. 2.5], between the wall *AB* and the extremity of the beam *C*, namely, when the gravity of portion *FGCF*, balanced from the point of rest *G*, has a greater ratio to the resistance in *FG* than the gravity of the whole beam *BAC* from the point of rest *AD* has to the resistance in *AB*.

But it is asked what sort of line must *BFC* be so that the resistances are proportional to the corresponding gravities, and the beam resists equally everywhere: it will be discovered that this line is parabolic. Indeed, the

[6] Here Leibniz is using the result that the concave area between the parabola and its axis is a third of the area of the square. This follows straightforwardly by his calculus, since the area is the integral of ydx between 0 and 1 with $y = x^2$, giving 1/3.

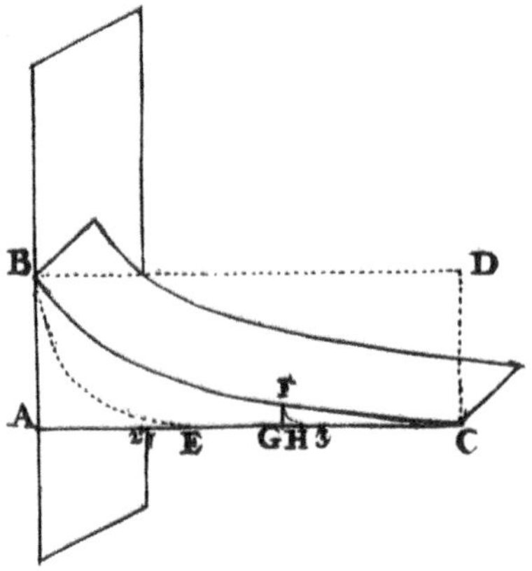

Fig. 2.5

resistance in *FG* is to the resistance in *BA* as the three-sided concave parabolic figure *FGHF* is to another *BAEB*, if the base of the three-sided figure is equal to its height (as is clear from the preceding): that is, as the square *FG* is to the square *BA* (because such a three-sided figure is a third of the circumscribed square).

But the moment or gravity of any portion whatsoever of *FGCF* balanced from G is also to the moment of the whole beam *BACB* balanced from *A*, as the square of *FG* is to the square of *BA*, which is easily demonstrated from the nature of a parabola (since the portions *CGFC* and *CABC* are as the cubes of *CG*, *CA*). Furthermore, let *G3* and *A2* be quarter parts of these (CG and *CA*), and they will be the distances of the centres of gravity of the portions *CGFC* and *CABC* from points of rest or centres of balance *G* and *A*, [323/324] and the moments of the said proportions will be as the products of the portions and the distances, that is, in the compound ratio of the portions or cubes of *CG* and *CA* and of the distances, which are as *CG* and *CA* (and therefore in the ratio of the squares of the squares of *CG*, *CA*, that is, in the ratio of the squares of *FG* and *BA*). Therefore, the resistances are proportional to moments or forces, that is, the proportion of each moment to its resistance will be the same everywhere, and indeed, the strength by which the beam resists its own weight will also be the same everywhere: and hence a beam configured in this way, extended to any length, if it is not broken by its own weight near the wall, will not break anywhere else either.

Moreover, since the prismatic parabolic beam *CABC* is only a third of the full beam *CDBA*, the parabolic beam will be six times stronger than the full

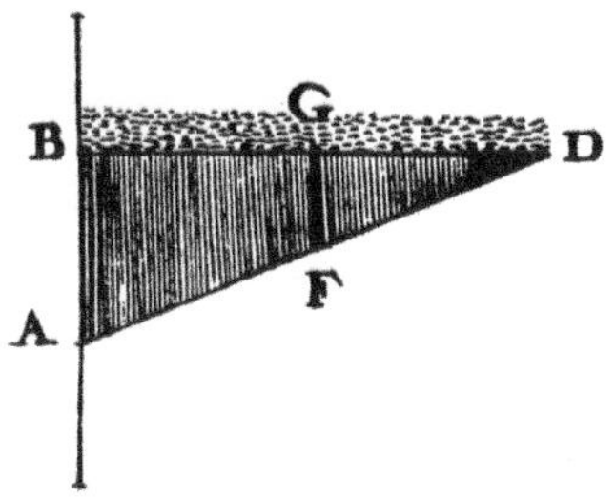

Fig. 2.6

beam, having taken only a third of the weight from that source, and having drawn its centre of gravity from *AG* to its half *A2*.

Setting aside the weight of the beam, let it be understood that the force of water or wind, or of something else is distributed equally along the whole length of the beam, as in *fig.* 6 [Fig. 2.6]. The joist *ABD* jutting out from the wall should bear the burden of the heaped earth, or grain, or other matter, it could be triangular, and the line *AD* could be straight, and the joist will equally resist the imposed weight everywhere, so that if it should not to break at the wall, it would not be able to break anywhere else: for from known laws of mechanics, the momentum of the pressing weight to *GD* is to the momentum of the pressing weight to *BD*, as *GD* squared to *BD* squared, or as *GF* squared to *BA* squared; that is, as the resistance in *GF* to the resistance in *BA*: but if it is supposed that the weight is in part imposed, in part from the figure of the beam, I can nonetheless render the resisting figure in the same way.

Thus far we have only considered a beam whose surface, by which it adheres to a wall or support, is the same height everywhere, which is why it was sufficient to assume that *BA* is straight. But since the common surface of the beam and the wall could vary, let us give a general rule for determining its resistance geometrically; if someone has the leisure to work out its special cases, he will discover many very elegant theorems. In general, however, let there be a beam *ABHC* (*fig.* 7 [Fig. 2.7]), whose section at the support *DE* is a plane *ABH* of any shape whatsoever. Let that [plane] be lowered horizontally, that is, let another plane *AGH* that is equal, similar, and similarly placed to it be described in the horizontal plane. Let a perpendicular line *GF* (equal to *BF*) be drawn to *AH* from point *G* (which corresponds to point *B*), and which is as far as possible from *AH*, the lowest of the horizontals; and let there be a cylindrical body, whose base or a certain section [324/325] parallel to the horizontal plane is similar and equal to *AGH*, but whose height is the

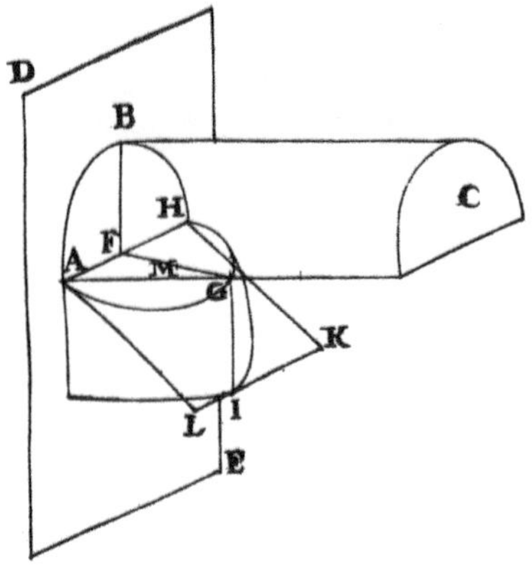

Fig. 2.7

perpendicular line *GI*, equal to *FG* or *BF*; we may call that body a cylinder. Let an indefinite tangent *KIL* be drawn parallel through *I* to *AH*. Finally, let a plane pass through *AH* and *KL*, which will make a half-right angle to the horizontal and will cut the cylindrical body in two parts, then the part on which *GI* falls, which in the figure is above the dividing plane, is called an 'ungula' by geometers. I say that this ungula, cut from the cylinder and serving as a lever whose fulcrum is *AH*, is equal to or represents the transverse resistance to breaking at *AHB*, if the weight of the cylinder is sufficient for tearing the same beam directly from the wall. But so that there is no need to consider the ungula as a lever, and so that we have a weight representing resistance absolutely, let the ungula be suspended from the point *M*, that is, from *FM*, the distance from the wall to the centre of gravity of the ungula; and in this way it will exactly equal the transverse resistance, if the whole cylinder equals the direct resistance.

Thus, if it is asked whether and where a solid object must break, an estimation will not be difficult for a geometer; for either that will not happen or it will happen chiefly at that place where the moment of the ungula (ôr what is made from having drawn the ungula to its centre of gravity, from the vertical plane in which lies the axis of the balance) will have the least ratio of all to the power trying to break it there: so that, from these few considerations, this whole matter is reduced to pure geometry, which is especially desired in physics and mechanics.

An addition: if anyone should seek some convex solid of equal resistance to this, a parabolic tube will suffice. In *fig*. 8 [Fig. 2.8], let there be a parabolic line *AEC* whose vertex is *A*, a line *AB* tangent at the vertex, around which the

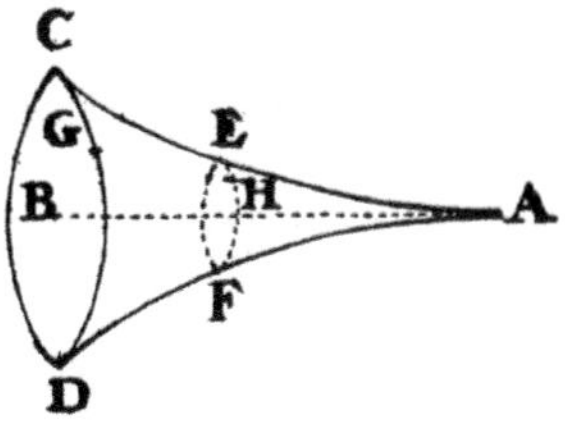

Fig. 2.8

parabolic line is rotated as if around an axis, and the tube *AECGDFA* will be made. If we now take another portion of the tube *AEHFA*, then, since the resistance of the bases ôr circles *CGD* and *EHF* are as the cubes of the diameters *CD* and *EF*, it will be found, from the nature of the parabola, that the moments of the portions *AECGDFA* and *AEHFA* are also as the cubes of *CD* and *EF*.

3

A Brief Demonstration of a Notable Error by Descartes and others concerning a Law of Nature[1] according to which they maintain that God always conserves the same Quantity of Motion in matter, a law which they also misuse in mechanics

G. W. Leibniz, *Acta eruditorum*, March 1686[2]

Several mathematicians, seeing that speed and magnitude[3] compensate one another in the five common machines, estimate motive force generally by quantity of motion, i.e. by the product of the body multiplied by its speed. Or, to speak more geometrically, they say that the forces of two bodies (of the same kind) set in motion, acting equally by their magnitude and by their motion, are in the compound ratio of the bodies or magnitudes and of their velocities. So,

[1] From the Latin '*Brevis Demonstratio Erroris memorabilis Cartesii & aliorum circa legem naturae, secundum quam volunt a Deo eandem semper quantitatem motus conservari; qua & in re mechanica abutuntur*', *Acta eruditorum*, March 1686, 161–3 (French translation, *Nouvelles de la République des Lettres* 1686, 996–9). (ESP 90–93; Gallica 256–58; GM VI 117–19; and A VI 4, 2027–30 from a draft.) Translated by Richard T. W. Arthur.

[2] Leibniz sent this paper to Otto Mencke, the editor of the *Acta eruditorum*, in a letter dated 6 January 1686. Mencke's reply (A II, 1, N. 473) is dated 27 January 1686. This is the article in which Leibniz publicly launched his challenge to the physics of the Cartesians. It was directed against the assumption that the motive force that is conserved in collisions is given by the Cartesian measure of 'quantity of motion', the product of the magnitude of a body and its speed. In its place Leibniz advocates an estimation of force as proportional to the effect it can produce, such as the height through which the same body can be raised under its own weight. By a deceptively simple reductio argument, Leibniz shows that the Cartesian measure is in conflict with Galileo's results concerning the speed acquired by a body falling under gravity.

[3] Here we are translating Leibniz's *moles* [bulk] as 'magnitude', and *celeritas* [swiftness] as 'speed'. This matches the French translations of them as *grandeur* and *vitesse*.

Leibniz: Journal Articles on Natural Philosophy. Richard T. W. Arthur, Oxford University Press.
DOI: 10.1093/oso/9780192843531.003.0004

since it is consistent with reason that the same sum of motive power be conserved in nature—neither being diminished, since we see no force being lost by one body without its being transferred to another, nor increased, since for this reason a mechanical perpetual motion never happens, because no machine, and therefore not even the entire world, can increase its force without a new impulse from without—this led Descartes, who held *motive force* and *quantity of motion* to be equivalent, to proclaim that the same quantity of motion is conserved in the world by God.[4]

In order to show what a great difference there is between these two concepts, however, I suppose, *first*, that a body falling from a certain height acquires a force [161/162] for rising back up again to the same height, if its direction carries it that way and there are no external impediments. For example, a pendulum will return precisely to the height from which it is dropped, unless the resistance of the air and other similar tiny impediments absorb something of its force, which we shall now refrain from considering. I also suppose, *secondly*, that the same amount of force is needed to lift a body *A* of one pound to a height *CD* of four metres[5] as is needed to lift a body *B* of four pounds to a height *EF* of one metre [Figs 3.1, 3.2].

All these things are conceded by the Cartesians as well as by the other philosophers and mathematicians of our times. Hence it follows that the body *A*, let fall from a height of *CD*, would acquire precisely as much force as the body *B* falling from a height of *EF*. For the body (*A*), after it has fallen from *C* and reached *D*, there has the force to rise up again to *C*, *by supposition 1*; that is, the force to lift a body of one pound (namely, its own body) to a height of four metres. And similarly, the body (*B*), having fallen from *E* to *F*, there has the force to rise up again to *E*, *by supposition 1*; that is, the force to lift a body of four pounds (namely, its own body) to a height of one metre. Therefore, *by*

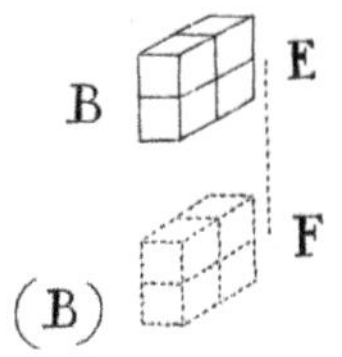

Fig. 3.1

[4] See Descartes, *Principia Philosophiae*, II, 36 (AT VIIIA, 61 ff.).

[5] Here we have substituted 'metre' for the measure Leibniz used of an *ell* [*ulna*, French *aune*]—a local measure of varying length, but in this case some 1.18 to 1.20 metres.

Fig. 3.2

supposition 2, the force of the body (*A*) when it is at *D*, and the force of the body (*B*) when it is at *F*, are equal.

Now let us see whether the quantity of motion is also the same in the one case as in the other. In fact, a very great difference will be found, contrary to expectation. Which I show as follows. It has been demonstrated by Galileo that the speed acquired in the fall *CD* is twice the speed acquired in the fall *EF*.[6] So if we multiply the body *A* which is as 1, by its speed which is as 2, the product, or quantity of motion, will be as 2; again, if we multiply the body *B* which is as 4, by its speed which is as 1, the product, or quantity of motion, will be 4. Therefore, the quantity of motion the body (*A*) has when it is at *D*, is half the quantity of motion that (*B*) has when it is at *F*. Thus, there is a great difference between motive force and quantity of motion, so that one cannot be estimated by the other, as we undertook to show. From this it is apparent how force is to be estimated by the quantity of the effect that it can produce; for example, by the height to which a heavy body of a given magnitude and kind can be lifted, but not, indeed, by the speed which it can impress on the body. For, [162/163] in order to give the same body double the speed, a force that is not double, but more than double, is needed. No one should be surprised that in the common machines—the lever, the windlass, the pulley, the wedge, and the screw, and the like—there is an equilibrium, since the magnitude of one body is compensated by the speed of the other, a fact that has its origin in the disposition of the machine; that is to say, when the magnitudes (supposing bodies of the

[6] Galileo Galilei (1638, III, Theor. II, Prop. II., 209).

same kind) are reciprocal to the speeds, the same quantity of motion is produced on either side. For in that case it happens that the quantity of effect, or height of descent or ascent, will be the same on either side, whichever side of the balance you wish to set in motion. Thus, in this case it is accidental that force can be estimated by quantity of motion. But other cases can be given, such as the one supplied above, where they do not coincide.

Meanwhile, since nothing is simpler than our proof, it is surprising that it did not occur to Descartes or the Cartesians, who are very learned men. But the former was led astray by too great a faith in his own genius, the latter, in someone else's. For Descartes, with a vice typical in great men, finally became a little presumptuous, while I fear that not a few Cartesians are gradually beginning to imitate many of the Peripatetics, whom they mocked; that is to say, they are becoming accustomed to consulting the books of their master instead of correct reason and the nature of things.

It must be said, therefore, that forces are in the compound ratio of bodies (of the same specific gravity or solidity) and the heights producing the speed, that is, the heights from which they could acquire such speeds by falling; or, more generally (since sometimes no speed may yet have been produced), the heights that would be produced; but not generally in proportion to the speeds of the bodies, however plausible this seems at first sight, and has seemed to many people. From the latter a great many errors have arisen, which can be discerned in the mathematico-mechanical writings of the reverend fathers Honoré Fabri and Claude Dechales, and likewise of Giovanni Alfonso Borelli,[7] and of other men who have otherwise distinguished themselves in these studies. Moreover, I believe that it is as a result of this error that some learned men have recently called into doubt Huygens' Rule concerning the centre of oscillation of pendulums, which is completely true.[8]

* * * * *

A French translation of this was published as 'Demonstration courte d'une erreur considerable de M. Descartes & de quelques autres touchant une loi de

[7] See Honoré Fabri, *Tractatus Physicus De Motu Locali* (1646); Claude Dechales, *Cursus seu mundus mathematicus* (1674); Giovanni Alfonso Borelli, *De vi percussionis* (1667) and *De motionibus naturalibus a gravitate pendentibus* (1670). For biographical information of these three thinkers, see the Introduction.

[8] Part IV of Huygens' *Horologium oscillatorium* (Huygens 1673) is devoted to finding the centres of oscillation of various figures. Leibniz is referring to propositions such as Part IV, Prop. III: 'If any number of bodies fall or rise, but through unequal distances, the sum of the products of the height of the descent or ascent of each, multiplied by its corresponding magnitude, is equal to the product of the centre of gravity of all the bodies multiplied by the sum of their magnitudes' (Proposition III; Huygens 1986, 112).

la nature selon laquelle ils soutiennent que Dieu conserve toujours dans la matiére la même quantité de mouvment, de quoi ils abusent même dans la mechanique. Par G. G. L.' *in the* Nouvelles de la Rèpublique *of September 1686 (995–999), immediately preceding the Abbé de Catelan's reply. It was prefaced by the following note:*

> *We have received from Paris the response that a learned Cartesian has made to an objection of Mr Leibniz, and we are assured that the curious will be pleased to see it here. Since this objection appeared in the month of March in the* Acta Eruditorum, *we could have dispensed with putting it in these* Nouvelles; *however, we put it here translated into French without having regard to the need to lay the groundwork. We put it here, I say, as much because it has been published only in Latin, as because not all those who see the Response here will have the convenience of consulting the Journal of Leipzig at the same time.*

4

A Brief Comment by the Abbé D. C., showing Mr G. G. Leibniz the paralogism contained in the preceding objection[1]

François Abbé de Catelan, *Nouvelles de la République des Lettres*, September 1686[2]

Mr Leibniz is astonished[3] that his proof, which he thinks is the simplest in the world, did not occur to the mind of Mr Descartes, or to the minds of the Cartesians. But it would be far more astonishing if so perceptive a philosopher and geometer had been able to fall into such an idea through carelessness, and to drag so many able people along with him. Let the learned world judge whether it is he or Mr Leibniz who has gone astray through too great a confidence in his own intelligence, the common fault of great men. Mr Leibniz raises a concern which is indeed that of an able mind, but he is rather too wide of the mark when he fears that the disciples of Mr Descartes are imitating the Peripatetics they deride. Let us take a look at this 'serious error' he is aiming to eliminate.

[1] From the French: '*Courte Remarque de M. l'Abbé D. C où l'on montre à M. G. G. Leibnits la paralogisme contenuë dans l'objection précédente*', *Nouvelles de la République des Lettres* 1686, 1000–4. Translated by Richard Francks, on the basis of a previous joint translation with Roger Woolhouse. (GP III 40–42; Gallica 257–8).

[2] This is a reply to Leibniz's '*G.G.L. Brevis Demonstratio Erroris memorabilis Cartesii et aliorum circa Legem naturalem, secundum quam volunt a Deo eandem semper quantitatem motus conservari, qua in re mechanica abutuntur*' (*Acta Eruditorum* 1686, 161–3). A French translation was included in the September *Nouvelles* immediately before this article (1686, 996–9). Parts of Catelan's article are quotations from or close paraphrases of that translation, although not marked as such. In the article Catelan offers a refutation on behalf of the Cartesians of Leibniz's argument from the preceding paper. He alleges that Leibniz has failed to realize that the Cartesian measure of force concerns *isochronous* powers, or motions impressed in equal times.

[3] This follows the French translation of Leibniz's expression as 'on doit s'étonner' [one should be astonished], which is more forceful than Leibniz's Latin '*mirum est*' [it is a wonder, it is surprising].

Leibniz: Journal Articles on Natural Philosophy. Richard T. W. Arthur, Oxford University Press.
DOI: 10.1093/oso/9780192843531.003.0005

He says:

1. that Mr Descartes asserts that God conserves in the universe the same quantity of motion.
2. that this same philosopher takes motive force[4] and quantity of motion as equivalent things.
3. that many mathematicians standardly estimate moving force by quantity of motion, or by the product of the body multiplied by its speed.

He then claims that these things are inconsistent with one another, that therefore moving [1000/1001] force and quantity of motion are very different, and that this rule of Mr Descartes's is false: *that the same quantity of motion is always conserved in nature.*

As far as the last part of his conclusion is concerned, it is up to Cartesian readers to consider how it might connect with *his premises*. As for the first, he proves it like this:

> According to Mr Descartes and other mathematicians, no less force is needed to raise a body of one pound to a height of four metres[5] than to raise a body of four pounds to a height of 1 metre. From which it follows that the unitary body[6] falling from the quadruple height acquires exactly the same force as the quadruple body falling from the unitary height; for each of them would acquire a force such that if external obstacles were removed it would be able rise back up to where it had fallen from. Moreover, Galileo has shown that the speed which a body acquires in falling from a height of four metres is double the speed it acquires in falling from a height of one. So multiplying the one-pound body by its speed, that is to say 1 by 2, the product, or the quantity of motion, will be 2; and multiplying the four-pound body by its speed, that is to say 4 by 1, the product, or the quantity of motion, will be [1001/1002] 4. So, one of these quantities is half the other, even though shortly before the forces had been found to be equal—the forces, I say, which Mr Descartes does not distinguish from the quantities of motion. Therefore, etc.

[4] 'Motive force' translates *force mortice*, and 'moving force' translates *force mouvante*.

[5] As noted, the French '*aune*' (used by the French translator and Catelan) translates Leibniz's *ulna*, about 1.18 to 1.20 metres, for which we have substituted 'metre'.

[6] This paragraph is a close paraphrase of the French translation of Leibniz's *Brevis Demonstratio*, although 'unitary' [*le simple*] and 'quadruple' [*le quadruple*] are Catelan's own abbreviations for Leibniz's *un corps d'une livre*, and *la hauteur de quatre aunes*.

I am surprised that Mr Leibniz has not noticed the paralogism in this proof, for where is the man with any ability in mechanics who does not understand that the principle of the Cartesians with regard to the 5 simple machines[7] concerns *isochronous* powers, or motions impressed in equal times, when we compare two weights together? For it is demonstrated in the *Elements* that two moving objects which are unequal in volume, such as 1 and 4, but equal in quantity of motion, such as 4, have speeds which are in inverse proportion to their masses, such as 4 and 1; and as a result in equal times they always cover distances proportional to those speeds. Moreover, Galileo shows that the distances traced out by falling bodies are in the same ratio to each other as the squares of the times. Thus in Mr Leibniz's example, the one-pound body would rise to a height of 4 metres in a time of 2, and the four-pound body would rise to a height of one metre in a time of 1. So, because the times are unequal, it is not [1002/1003] strange that he should find the quantities of motion in this fall to be unequal—although they would have been found to be equal in a fall where equality in the times made it quite different from this one. Now let us suppose that these two bodies move only in the same times, i.e. that they are suspended on the same balance, and at distances inversely proportional to their sizes. We will find the quantity of their opposing motions, or the forces of their weights, to be equal, whether we multiply their masses by their distances, or by their speeds. Things happen differently when the times are unequal. From which it is clear that neither Mr Descartes nor anyone else is mistaken here, and I very much doubt that any of the learned men who have recently contested Mr Huygens' rule concerning the centre of oscillation will change their opinion because of this objection by Mr Leibniz.

As the original of this reply was written in Latin, and put into French by a man who may perhaps not always have understood clearly what he was translating, we would ask those who might want to reply to take good care that there is no misunderstanding which should be attributed to the translator. We will quickly make the position clear if necessary.

[*The remainder of the article (1004–05) is a summary by the Editor of a response from Denis Papin to a reply to his objections against a design for a perpetual motion machine.—ed.*]

[7] As noted by Leibniz in the *Brief Demonstration*, the five simple machines of antiquity were the lever, winch (or windlass, i.e. wheel and axle), pulley, wedge, and screw.

5

A Reply by Mr L. to the Abbé D. C., contained in a letter written to the Editor of these *Nouvelles* on 9th January 1687, concerning what was said by Mr Descartes: that God always conserves in nature the same quantity of motion[1]

G. W. Leibniz, *Nouvelles de la République des Lettres*, February 1687[2]

Having found in your *Nouvelles* for the month of September last my objection against the famous Cartesian principle concerning the quantity of motion, with the response from a Cartesian savant in Paris named [131/132] the Abbé C., I am sending you my reply in order that all the papers in the case should appear together, if you think it appropriate. In fact, this is intended only to clarify the matter by pursuing rather than justifying the objection I raised, because the Abbé did not actually bring anything against it, and he grants me more than I need. But I am very concerned that the other Cartesians will disown him. According to him their rule is reduced to a very minor thing,

[1] From the French: '*Replique de M. L. à M. l'Abbé D. C. contenuë dans une lettre ecrite à l'Auteur de ces Nouvelles le 9 Janv. 1687. Touchant ce qu'a dit M. Descartes que Dieu conserve toujours dans la nature la même quantité de mouvement.*' *Nouvelles de la République des Lettres*, February 1687, 131–44. (ESP 112–126; Gallica 414–417). Translated by Richard Francks, on the basis of a previous joint translation with Roger Woolhouse.

[2] In making this reply to the Abbé De Catelan, Leibniz would have been uncertain of the Abbé's status as a spokesman for the Cartesian cause. But he would have assumed that his answer would be read by Malebranche, with whom, as noted in the Introduction, Catelan was closely associated. Accordingly, since Malebranche had made some criticisms of Descartes's rules of collision, Leibniz uses this opportunity to attempt to win Malebranche over to his side.

Leibniz: Journal Articles on Natural Philosophy. Richard T. W. Arthur, Oxford University Press.
DOI: 10.1093/oso/9780192843531.003.0006

since he claims that it is only *a special principle with regard to the five simple machines, which concerns isochronous powers, or motions impressed in equal times.*

I had shown that in a certain quite common case, and in an infinity of other similar ones, two bodies have the same force even though they do not have the same quantity of motion. He accepts that, and that is all I claim. But he adds that we should not be surprised by it, because in the case proposed the two bodies have acquired their forces in unequal times—as if this principle must be restricted to those which have been acquired in equal times. That means he concedes me the case, and that is more than I ask. But I would be wrong to want to claim an advantage against the Cartesians from the fact that someone should defend them in this way, because I do not think the Abbé C. will ever find one, at least among those [132/133] who can be called Geometers, who accepts his restriction. Cartesians claim in general that there is maintained the same total force, which they always determine by the quantity of motion, or the sum of the products of the masses multiplied by their speeds. For example, if there is a body of 4 pounds with one degree of speed, and we suppose that all its force is now to be transferred into a body of one pound, is it not true that the Cartesians will declare that given this hypothesis the body must receive a speed of four degrees, in order that the same quantity of motion should be preserved? For a mass of 4 multiplied by a speed of 1 produces the same as a mass of 1 by a speed of 4. But according to me the body must receive only a speed of 2 (as I shall prove later), so the disagreement is quite clear. And in estimating in this way the forces the bodies have received, neither these gentlemen nor any others that I know of, apart from the Abbé C., give any consideration to whether those forces have been acquired in times that are long or short, equal or unequal. In fact time plays no part in determining this. Seeing a body of a given size travelling with a given speed, would we not be able to determine its force without knowing in what time, and with what detours and delays, it has acquired the speed that it has? It seems to me that we can decide here about the present state [133/134] without knowing the past one. When there are 2 perfectly equal and similar bodies, which have the same speed, but acquired in one of them by a sudden impact, and in the other by a descent of a considerable duration, will we say that their forces are different? That would be like saying that a man is richer if his money took him longer to earn. But what is more, it is not even necessary that the two bodies I postulated should have covered their different heights in unequal times, as the Abbé C. imagines, because he has not realized that we can change the time of the descent however we like by changing the line of descent, making it more or less inclined, and that there is an infinity of ways in which we can bring it about that the two bodies descend from their different heights in equal times. For, abstracting away from the resistance of the air and similar impediments, we can make a body descending from a given height acquire the same speed, whether the descent is perpendicular and fast, or inclined and slower. And so it follows that a difference in times has no effect on my objection.

These things are so obvious that I would perhaps have grounds for throwing back at the Abbé C. certain expressions that he uses, but I think it better not to amuse oneself in that way. Indeed, I think that because my objection is [134/135] so simple, that very fact has served to mislead him, since it did not seem to him credible that such a straightforward point might have escaped so many clever people. That is why when he noticed the difference in times he seized upon it, without giving himself a chance to see that it is merely accidental. Now I have a sufficiently good opinion of his mind and of his sincerity to hope that he will recognize this himself, and I think that what follows will go even further in showing him what is involved.

Also, in order to avoid the doubts of those who might think they can resolve my objection by saying that the imperceptible matter which presses down heavy bodies and produces their acceleration has lost exactly the quantity of motion that it gives to the bodies, I reply that I agree that this pressure is the cause of heaviness, and I think that this aether does lose as much force (but not as much motion) as it gives to heavy bodies. But all that does nothing to resolve my objection, even if I were to accept (as I do not) that the aether lost as much motion as it gave. For my objection is deliberately formulated in such a way that it is not at all important how the force has been acquired, which is something I have abstracted from, so as not to get into a dispute about any particular hypothesis. I take the force and the speed acquired just as they are, without at the moment concerning myself [135/136] as to whether they were given all at once by the sudden impact of another body, or little by little by a continual acceleration due to heaviness, or to elasticity. It is enough for me that the body now has that force, or that speed. And then I show that its force must not be determined by the speed or quantity of motion, and that the body can give its force to another without giving it its quantity of motion, and therefore that when this transfer happens it can, and indeed must, come about that the quantity of motion diminishes or increases in the bodies, while the same force remains.

I shall therefore now prove what I had proposed above, that *in the case where we suppose that the whole force of a body of 4 pounds, whose speed* (which it possesses, for example, when travelling in a horizontal plane, however it may have been acquired) *is one degree, is to be given to a body of one pound, then this latter will receive not a speed of 4 degrees, in accordance with the Cartesian principle, but of two degrees only*, because in this way the bodies or weights will be in inverse proportion to the heights to which they can rise in virtue of the speeds they have; and those heights are as the squares of the speeds. And if the four-pound body of 4 pounds, having a speed of one *degree* in a horizontal plane, on getting picked up [136/137] by colliding with the end of a pendulum or perpendicular wire, rises up to a height of one foot, then the one-pound body will have a speed of two degrees, so that, if it were picked up in the same way, it would be able to rise to 4 feet. For it takes the same force to raise 4 pounds to one foot and one pound to 4 feet. But if this one-pound body is to receive 4 degrees of speed, as per Descartes, it would be able to rise to a height of 16 feet, and as a result the same force which could raise four pounds to one foot, transferred into one pound, would be able to raise it to sixteen feet. Which is impossible, because the effect is four times as large, so we would have gained, and pulled out of nowhere, three times the force there was before. That is why I think that instead of the Cartesian principle we could establish a different *law of nature* which I hold is the most universal and the most inviolable, namely *that there is always a perfect equation between the complete cause and the complete effect.* This law does not say only that effects are proportional to causes, but in addition that every complete effect is equivalent to its cause. And although this axiom is very metaphysical, it is still one of the most useful that we can employ in physics, and it gives a way of reducing the forces to a geometrical calculus. But to show more clearly how it should be used, and why Descartes and others have [137/138] strayed so far from it, let us consider his third rule of motion as an example. Let us suppose that two bodies, B and C, each of one pound, run into each other, B with a speed of one hundred units, C with a speed of one unit. Their total motion will be 101. But if C with its speed could rise to one inch in height, B with its speed will be able to rise to 10,000 inches, so the force of the two together will be enough to raise one pound to 10,001 inches. Now, according to this *Third Cartesian Rule*, after the collision they will both go along together with a speed of 50½, so that when we multiply it by two (the number of pounds that go along together after the impact) it comes to the initial quantity of motion of 101. But then these two pounds will only be able to rise together to a height of 2,550¼ inches (which is the square of 50½), which is the same as if they had the force to raise one pound to 5,100½; whereas before the impact there was enough force to raise one pound to 10,001 inches. Therefore, nearly half the force will be lost, according to this rule, with no explanation, and not having been used for anything. Which is as little possible as what we showed earlier in the other case, where in virtue of the same general Cartesian principle we could gain three times the force with no explanation. [138/139]

The famous author of the *Recherche de la Vérité*[3] has rightly seen some errors by Mr Descartes in these matters, but as he was presupposing the principle which I am rejecting, he thought that of the seven Cartesian rules the first, second, third, and fifth were true, when only the first, which is obvious in itself, is defensible. The same author of the *Recherche*, reasoning on the supposition of hard bodies with no elasticity, claims that they only have to rebound or separate from each other after impact if they come together with speeds in inverse proportion to their sizes, and that in all other cases they will go along together after the impact, maintaining the original quantity of motion. But here is a big problem I find in this. Let body *B* be 2, with speed 1, and body *C* 1, with speed 2, which run directly into each other. He accepts that they will rebound with the speeds they had. But if we suppose the speed or size of one of the bodies, say, *B*, to be increased by however small an amount, he claims they will both go along together in the direction in which *B* alone was originally travelling—and that will be with a speed of almost four thirds, if we suppose the change made in respect of *B* is so small that in calculating the quantity of motion we can keep the original numbers without any significant error. But is it credible that for a change as small as you like in [139/140] the supposition with regard to body *B* there should result such a great difference in the outcome, such that all the rebound ceases, and *B*, which originally had to move back with a speed of 1, now, as a result of having however tiny an amount more force, must not only not go backwards, but must actually go forwards with a speed of almost four thirds? Which is even more strange when before the impact it was only moving forward with a speed of virtually 1. So by running into it, the other body, instead of making this one go backwards, or go forward less, would make it go forward more, as if it were pulling it towards it—which is completely implausible. As it is the author of the *Recherche de la Vérité* to whom we are beholden for the correction of some rather serious Cartesian prejudices both elsewhere and on this subject, it seemed to me appropriate to show here what still remained to be said. And since I am confident that he has no less integrity than acuity, far from being afraid that he will be offended, I am hoping for his approval.

However, I think that Mr Descartes, who forgot in his rules to note the case when two unequal bodies run into each other with unequal speeds, would have

[3] The author is Nicolas Malebranche: see the Introduction for explanation.

been obliged in the case above to say the same thing as the author of the *Recherche*, [140/141] as far as I can tell by Rule 3, on which both are agreed. But there too we will find the inequality of effect and cause, as it would be easy to show by calculation following the example of the third rule. This inequality is also found in what the author of the *Recherche* says to correct Mr Descartes's fourth, sixth, and seventh rules. For example, as regards the sixth, let B be one pound, speed 4, C one pound, at rest. He claims that they will go along together after the impact with a speed of 2. Therefore, whereas before there was a force capable of raising one pound to sixteen feet, now there will only be a force capable of raising 2 pounds to four feet, and half the force will have been lost. According to Mr Descartes in this case *B* and *C* will go in the same direction, and the speed of *B* will be 3, that of *C* 1. Therefore, in total there will be a force capable of raising one pound to 10 feet, and more than a third of the force will be lost.

What might have led such excellent writers astray, and what has further confused this question, is that we have seen that bodies whose speeds are in inverse proportion to their extensions stop one another, whether in a balance, or not in a balance. That is why it was thought that their forces were equal, *the more so because they did not observe anything in the bodies other than their speed and extension.* But this is where [141/142] they could usefully have employed the distinction which exists between force and direction, or rather between the absolute force which is necessary to produce some continuing effect (for example to raise a certain weight to a certain height, or to tighten a certain spring to a certain extent), and the force to advance in a certain direction, or to maintain a direction. Because although a body of 2 with speed 1 and a body of 1 with speed 2 stop one another, or prevent one another from advancing, nevertheless if the 1 can raise a pound to 2 feet in height, the 2 will be able to raise 1 pound to 4 feet in height. This is paradoxical, but undeniable after what we have just said. We could however give a new interpretation to the principle of the quantity of motion, and after this correction it would remain universal; but it is not easy to see it.

I will add a remark of some importance for metaphysics. I have shown that force should be calculated not by the combination of speed and size, but by future effect. However, it seems that force or power is something real in the present, and future effect is not. From which it follows *that we will have to accept in bodies something different from size and speed, lest we want to refuse to bodies all power of action.* I think moreover [142/143] that we have not yet even fully understood matter and extension. The author of the *Recherche de la Vérité* has recognized this obscurity with regard to the soul and thought, as against the usual opinion of Cartesians, but with regard to matter and extension he seems to agree with them. However, there is an indicator for recognizing whether something is sufficiently well known, which I have set out in a little essay which can be found in the Journal of Leipzig for November 1684 on the abuse of ideas and of knowledge claimed to be clear and distinct, which I draw on now, as well as on what I have said here and there in the same journal, regarding the imperfection of Mr Descartes's Geometry and analysis.[4] I make mention of these things here so that it should not be thought that it is lightly and without some knowledge of the case that I have suggested that we should not content ourselves with paraphrasing Mr Descartes, and that those who follow that famous author (whose works I admire as they deserve) should go back over a number of passages in his writings to see how they stand up to reason and to nature—the more so since one of his most famous decrees, and one which seemed best established, has just now been annulled. I am confident that the truly able among those who are called Cartesians will not be angered by these remarks, and I [143/144] believe there are some who could give us something as fine as what Descartes himself gave us on for example salt, or the rainbow. It is perhaps only too great an attachment to the works of the master which prevents them. The sectarian spirit is naturally contrary to progress; to go forward we have to take things in a new light, which is not easy when our minds are too filled with borrowed ideas which have been accepted on the basis of authority more than of reason. I am, etc.

[*The remainder of the article is a note by the editor on the publication of a book on the life of the late Duke of Hanover, containing among other things a poetic eulogy by Leibniz—ed.*]

[4] Here Leibniz refers to his '*Meditationes de Cognitione, Veritate & Ideis*' of November 1684, his '*De dimensionibus figurarum inveniendis*' of May 1684, and his '*De geometria recondita*' of June 1686.

6

Comment by the Abbé D. C. on Mr L.'s Reply with regard to Mr Descartes's principle of mechanics; in Article 3 of these *Nouvelles*, February 1687[1]

François Abbé de Catelan, *Nouvelles de la République des Lettres*, June 1687[2]

Mons. L's Reply *[to the Editor]*:

> *"Having found in your* Nouvelles *for the month of September last my objection against the famous Cartesian principle with regard to the quantity of motion, with the response from a Cartesian savant in Paris named Mr [D.]*[3] *C., I am sending you my reply in order that all the papers in the case might appear together."*

Comment

I hereby declare that I am most obliged to Mr L. for the honour he does me in calling me a 'Cartesian savant'. But at the same time the truth does not permit me to conceal [577/578] the fact that I do not have sufficient talent or standing to be worthy of the title. It is more than enough for me, without claiming any rank among the 'sects of the savants', if I can adequately fill the minor place

[1] From the French: '*Remarque de M. l'Abbé D.C. sur la Replique de Mons. L. touchant le principe mecanique de Mons. Descartes, contenuë dans l'article III de ces Nouvelles, mois de Fevrier 1687.*' *Nouvelles de la République des Lettres*, June 1687, 577–90. (Gallica 522–25.) Translated by Richard Francks, on the basis of a previous joint translation with Roger Woolhouse.

[2] In this article De Catelan responds to Leibniz's arguments in his previous reply, arguing against Leibniz that the comparison of motions necessarily presupposes that the times of the motions should be equal, and denying any necessary connection of Descartes's mechanical principle with the principle of conservation of motive force.

[3] The text had 'M. L. C.', instead of 'M. D. C.', i.e. Mr De Catelan.

Leibniz: Journal Articles on Natural Philosophy. Richard T. W. Arthur, Oxford University Press.
DOI: 10.1093/oso/9780192843531.003.0007

which the Author of nature has given me in the School of *Reason*, which is open to all men, and where the front seats are reserved for those who make themselves most worthy of them, by applying their minds to the Search for Truth, and by working for its discovery; among whom it seems to me we can include Mr Descartes without being a professed 'Cartesian'. If from the first I have lined up on the side of that philosopher in Mechanics, it is because in dealing with it he seems to follow so well the lessons of that famous School, so that embracing his opinions in this case is only accepting the evidence, and submitting oneself to the judgement of reason. But I find myself here engaged in a suit; one does not plead one's case without stating one's position; here I must declare myself either Cartesian or Anti-Cartesian; I take the side of Descartes in Mechanics, so I am a Cartesian; so be it: I will not [578/579] refuse the name, provided that the word 'Cartesianism' and the word 'Reason' just mean the same thing. But let us get to the bottom of this suit we are concerned with. As I do not feel that I am very skilful in the art of evasion, I shall go straight to the point. That way I am less afraid of ruining my case.

When I examined Mr L.'s objection against Descartes, Borelli, the Reverend Fathers Fabri, and Dechales, and the other mathematicians who have written about Mechanics, I understood that he was claiming to prove that there is a contradiction between that philosopher's principle concerning Mechanics, the principle of Galileo concerning the fall of heavy bodies, and that of other writers on Mechanics concerning the measure of the quantity of motion. The first of those three principles, shown by an example in order to make it easier to understand, is *that it takes as much force to raise a body of one pound to a height of 4 feet as to raise one of four pounds to a height of 1 foot.* The second is *that the distances perpendicular to the horizon which heavy bodies cover in their fall towards the centre of the earth are as the squares of the time that the fall lasts.* The third is that *you get the proportion of the quantities of motion of two bodies if you multiply by the numbers which* [579/580] *express the ratio of their masses, the numbers giving the ratio of their speeds.*

The truth of these three propositions, if they are correctly understood, seems to me undeniable. I have been unable to agree with Mr L. on the contradiction he finds in them. In my opinion he confuses two very different things; that is what makes the conclusions he draws from them contradictory, and not that any of the principles is false. I will explain. The force of bodies considered in the bodies themselves and not in the will of the Prime Mover is what is called their *motion*, that is to say their *transfer from one place to another*, or their *continual change of situation with regard to the bodies which immediately surround them.* Their *relative force* is also nothing but their motion, but in so far as it is *communicable* to other bodies which are so placed as to oppose the direction of it. Motion is a mode or manner of being of *extension*, or *body*. Extension, when we consider it only under its three dimensions *length, breadth, and depth*, without paying attention to its ways of being, which are shape and motion, is called *space*. Since motion is the transfer of a body, it means that the body is continually applied under its three dimensions to the space within [580/581] which it is moved. All the space which the moving body occupies during its motion we can call the quantity of space covered; that to which a single one of the dimensions of the body is applied or corresponds according to the direction of the motion is called the length of space covered, or simply *the space covered*. So there is no motion unless there is some body moved, and some space covered. *Time* or *duration* relate to being, and to the modes of being—to *motion*, therefore, as well as to *body* or *extension*. As extension and time are divisible to infinity, so motion is capable of more and less to infinity. Several extensions taken together can be equal to a single one taken a certain number of times: two feet or two times one foot of extension are the same thing. Similarly, two times one unit of motion is as much as two units of motion. In all extension you can consider as many equal parts as you like; the motion of an extension or of body therefore belongs to all the parts which make it up, over which it is necessarily distributed in accordance with the proportion of the spaces that they cover together. Because of the divisibility of time, extension, and motion, our mind can, for example, distinguish indifferently [581/582] in the same body either 10 degrees of motion of the same length, distributed over 10 equal parts which compose the mass, or it can consider in the whole only a single motion of 10 degrees. According to the first way of conceiving the thing, where we have regard to the whole mass of the body moved, we say that a body has a certain *quantity of motion*; and according to the other, where we pay attention only to the transference, or the length of the space covered, we say that a body has a certain *speed*—but that is understood only by comparison of one body with another.

It is obvious from everything we have just said that a motion, of whatever kind it may be, of whatever body it may be, and whatever space it may have covered, has lasted a certain time. And if it had lasted longer, it would have covered a longer space; and if it had not lasted as long, it would have covered a shorter one. So, if we know that two equal bodies have covered two equal spaces, how do we ascertain whether they had equal or unequal motions? It is impossible, unless we know the duration of their motions, because if one has taken half an hour to move, and the other a quarter of an hour, we will certainly conclude that the motion of the second is double that of the first; and if the times have been equal, that the motions [582/583] have been equal as well. There are therefore encounters in which the consideration of time is necessary in order to be able to ascertain the moving forces; not that time adds anything to force, but because it enables us to know the ratio between the motions of which it is an attribute, as it is of all things that subsist.

Here is another example, which will show that the comparison of motions necessarily presupposes that of times. We know that two unequal bodies, one of which is four times the other, have equal motions, and we have to determine what spaces they cover in a straight line. It is clear that the quadruple body has only a quarter of the speed of the other one, because as it has no more motion, although it is four times as large, it must be that the quantity of that motion is shared between its four parts; otherwise, some of them would be at rest, which is contrary to our hypothesis. But all four move along together, because they make up a single whole which is moving in a straight line. They therefore have an equal speed, namely, *a quarter of the motion of the whole, to which quarter the motion of the simple body is equal. That is to say*, its speed, which here is the same thing. The four quarters of *the quadruple body*, or that entire body, therefore have only a quarter of the speed of the *simple* body. That is what we [583/584] know in this matter; but the ratio of the spaces covered can never be known as long as the duration of the motions is unspecified: if I choose to suppose that the quadruple body, which has only a quarter of the speed of the simple body, takes twice as long to move, I have no need of a proof to show that it will cover two quarters, or half, of the same space; if I make it move for four times as long it is no less clear that it will cover four times a quarter of the space which the other covers, that is to say, that it will go as far as that one does. If they move for the same time, or in equal times, the simple body, which we suppose to go four times as fast as the quadruple one, will make four times as much ground.

You can argue about this as much as you want; these things are so clear that in the end you have to accept them. So, when there is equality between the quantities of motion, or the forces, and inequality between the bodies moved, the spaces covered will not be reciprocal to the masses or proportional to the speeds unless there is still unity or equality of times; it does not matter if the forces increase or decrease, provided that the same happens to the similar parts—such as quarters, thirds, halves, etc.—of the spaces covered: in a word, provided that [584/585] the state of the question does not change.

I do not think that anyone could say to me here that he agrees with all this, but denies that I can conclude anything from it in favour of *the famous*

Cartesian Principle, against the objection of Mr L. Because I would ask him if he can really clearly understand that two bodies, one of which is *simple* and the other *quadruple*, are very different from two weights, one of which is of *one pound*, and the other of *four*; and that it is a quite different thing to *transport with an equal quantity of motion*—whatever it be—*the simple body* through a quadruple space and *the quadruple body* through the *simple* space, than it is to *raise with an equal force*—whatever force you please—*the weight of one pound* to *a height of four feet* and that *of four pounds* to a *height of one foot.* For myself, I see no difference. The reader who has examined carefully the arguments on the one side and on the other will judge who has lost and won between Mr L. and myself, and whether I have *seized upon the difference in times without giving myself the chance to see that it is merely incidental*,—as if one and the same motion does not have more or less effect depending on whether it lasts more or less time—it seems to me that the whole contradiction that Mr L. sees between the principle of Descartes and that of Galileo only comes from the fact that he contents himself with [585/586] judging the forces by their effects, or the motions by the spaces covered, without having any regard to their duration—which nevertheless really does produce, between those effects or spaces covered, a difference of more or less, which completely changes the question. Witness this example given by Galileo: two weights which fall as a result of their weight, one from a height of four feet and weighing one pound, and the other from a height of one foot and weighing four pounds, acquire speeds in inverse square relation to the spaces they cover from high to low, that is to say, of 2 and 1, and their quantities of motion, as I have explained above, are 2 and 4, or 1 and 2. So in this case the forces are unequal, whereas they are reckoned equal in the example of Descartes. But where is the contradiction? This philosopher is talking about moving forces applied in equal times, and Galileo compares forces applied, or motions acquired, in unequal times, and proportional to the square roots, 2 and 1, of their heights, 4 and 1. It is useless to reply as Mr L. does that on inclined planes, on which loads are made to rise or fall, the time is longer without changing the line of descent, because it is shown in Mechanics that the force needed to raise [586/587] a load on an inclined plane is less than that needed to raise it perpendicularly to the same height, the reason for which is that in the same time you raise it less high in proportion as the plane is longer, so that since the moving force is being applied a larger number of times, it produces the same effect even though it is smaller. And it does not matter that the weight descends on this plane as a result of its weight, because the force which produces its weight imprints it on it the more slowly as it has further along the plane to go in order to descend—which is the same thing as producing its weight in smaller parts, but repeated more often.

As for Principle 3, accepted up to now by all students of Mechanics, there is no problem with it given what has just been said. Multiplying the speeds 1 and 3 of two bodies *A* and *B* by their masses 4 and 5 is nothing different from multiplying together two numbers of which the second, 4 or 5, shows how many equal parts a body contains with respect to another; and the first, 1 or 3, expresses what ratio there is between certain motions which belong to each one of the equal parts of two bodies *A* and *B*, and which are called their speeds. So this product of the two numbers—4 and 1 on the one side, and 5 and 3 on the other [587/588]—is the ratio of each speed taken as many times as there are parts in each mass; that is to say it is the ratio of total motion, or the quantity of motion of each body, *A* and *B*, since the speed is the motion of the whole divided according to the number of equal parts, and the multiplication puts back together what the division has taken apart: 3 multiplied by 5 gives 15, just as 15 divided by 5 gives 3, and 1 multiplied by 4 gives 4, divided by 4 gives 1. So, to get the relative quantities of motion we have to multiply by the numbers which express the ratio of the masses the numbers showing the ratio of the speeds. I think I can assert here without being accused of vanity that if I am of the opinion of Descartes, of Borelli, of Fathers Fabri and Dechales, etc. on these kinds of question, it is not at all because I have let myself be *misled because it had not seemed to me credible that so straightforward a point as Mons. L.'s objection might have escaped so many clever people.* I say moreover, *I have now a sufficiently good opinion of his mind and of his sincerity to hope that he himself will recognize* that principles so evident and so certain have not thrown so many wise mathematicians into *a variety of errors* in Mechanics. [588/589]

[Leibniz's] Reply:

> *"And I think that what follows will help him to see what is involved even better."*[4]

[Catelan's] Comment:

I do not see an absolutely necessary connection between Descartes's mechanical principle and the fundamental proposition of his physics, that *God always conserves an equal quantity of motion in matter*, so that I would have to regard all the arguments that Mr L. gives in his response as so many reasons against

[4] The editors of the journal do not make clear where this purported quotation from Leibniz is from.

me. Even if the author of nature had established a law of the communication of motions according to which there were sometimes some which destroyed themselves or annihilated themselves, there would still be a necessity that there should be as much motion in a big body as in a small one, in order for them to cover in the same time spaces inversely proportional to their masses. Nevertheless, I am not unwilling to take sides on the rest of the dispute and to say what I think about the conservation of an equal quantity of motion in matter, and the rules of motion on the assumption of bodies that are hard in themselves. But I shall have to put that off for another time, because I see that my comment is already so long [589/590] that it is time to end it so as not to bore the reader further, and not to abuse the good nature of the Editor who is kind enough to include it in these *Nouvelles*.

7

Response by Mr L. to the Comment by the Abbé D. C. in Article 1 of these *Nouvelles* for the month of June 1687, in which he attempts to defend a Law of Nature proposed by Mr Descartes[1]

G. W. Leibniz, *Nouvelles de la République des Lettres*, September 1687[2]

We would have gone over a lot of ground again if the Abbé De C. had not frankly admitted in his Comment that he has not yet grasped my meaning. I do not understand on what basis he attributes to me a view which I have never considered; because he puts at the start three propositions, and says later (p. 580) that *he could never agree with me as to the contradiction I find in them.* But I, far from finding any contradiction in them, used them myself to show the error of the Cartesian principle. So, since nearly everything he says afterwards is intended only to explain and defend those propositions, it does not affect me at all. If misunderstandings like these arise in a dispute which is

[1] From the French: "*Réponse de M. L. à la Remarque de M. l'Abbé D. C. contenuë dans l'Article 1 de ces Nouvelles, mois de Juin 1687. où il prétend soûtenir une Loi de la Nature avancée par M. Descartes*, September 1687, 952–6. (ESP 137–141; Gallica, 622–3.) Translated by Richard Francks, on the basis of a previous joint translation with Roger Woolhouse.

[2] In this article Leibniz denies Catelan's charge that he had tried to show a contradiction between Galileo's Law of Fall and Descartes's mechanical principle, affirming that he had instead tried to show how together they contradict the Cartesians' identifying of quantity of motion as the force that is conserved in collisions. To drive home Catelan's deficient understanding of the non-uniform motion of a body falling under under gravity, and of equal increases of motion in equal times, Leibniz submits a challenge problem for the Cartesians, the problem of finding the isochronous line—the curve traced by a heavy body falling under gravity in such a way that its velocity vertically downwards remains constant.

Leibniz: Journal Articles on Natural Philosophy. Richard T. W. Arthur, Oxford University Press.
DOI: 10.1093/oso/9780192843531.003.0008

virtually about pure mathematics, what [952/953] can we expect in Morality, or in Metaphysics? This obliges me to beg the Abbé, and anyone else who hopes to be able to defend the Cartesians' principle, to reply clearly and point by point to what I had already said in the beginning, and which I will put here as numbered steps. Otherwise, we would be just wanting to amuse ourselves with irrelevances.

1. I ask whether it is not true that according to Mr Descartes a body of four pounds, the speed of which is unitary, has as much force as a body of one pound of which the speed is quadruple. So that if all the force of a body of four pounds is to be *transferred* to a body of one pound, that body must receive the quadruple of the former's speed, according to the Principle of the quantity of motion, on which Mr Descartes's Rules are based.
2. I ask also whether it is not true that if the first body, with one degree of speed, can lift four pounds (which is its weight) to one foot, or (what is equivalent) one pound to four feet, then the second, with four degrees of speed, will be able to lift one pound (which is its weight) to sixteen feet, in accordance with the demonstrations of Galilei and others? For bodies can rise to heights [953/954] which are as the squares of the speeds they have before rising.
3. And that thus it follows from the opinion of Mr Descartes that out of a force which could lift four pounds to one foot, or one pound to four feet, we will be able to make *by transference* a force capable of lifting one pound to 16 feet, which is its quadruple; and the surplus we will have gained, which is triple the original force, will have come out of nothing. Which is an obvious absurdity.
4. But according to me, and in fact, if all the force of a body of 4 pounds, whose speed is one degree, is to be transferred to a body of one pound, it would give it a speed of only two degrees, so that if the former could lift its weight of four pounds to one foot, the latter could lift its own weight of one pound to the height of four feet. Thus, *there is not the same quantity of motion retained; what is retained is the same quantity of force*—which is to be estimated by the effect it can produce.

It is rather surprising to see that no attempt has yet been made to reply clearly to things which are so precise and so straightforward, and that instead I have had attributed to me opinions that I do not hold. [954/955] However, *it would hardly be in keeping with the reputation for clear reasoning—on which the*

Cartesians pride themselves so much—if they showed so little of it when it came to defending one of their most famous principles. Would it not be said that since they could not defend their Master's error, they had tried at least to cover it up? And then what would become of those fine protestations of the love of truth, and that they aim to follow Descartes only where it is in keeping with reason? But I shall await something better.

I will add only, as an aside, that I grant to the Abbé that we can estimate the force by the time; but we must be careful about that. For example, we know the force a heavy body has acquired by the time it takes to attain it through falling, provided we know the line of descent. Because depending on whether it is more or less steep, the time will be different; whereas it is enough to know the height for working out the force which the body has acquired through descending from that height.

Now, this difference in times has made me think of a very nice *problem* which I have just now solved, and want to present here, in order that [955/956] our dispute should provide some occasion for the advancement of learning: *To find the line of descent along which a heavy body descends uniformly and approaches the horizon by equal amounts in equal times.*[3] The Analysis of the Cartesian gentlemen will perhaps give it easily.

[3] This is the famous challenge problem to find the isochronous curve, namely the curve that would be traced by a falling body which approaches the horizontal by equal amounts in equal times, so that it always maintains a constant velocity in its motion vertically downwards. Catelan did not rise to the challenge, as explained in the Introduction; but it gave Leibniz the opportunity to display his mathematical prowess to the embarrassment of the Cartesians. He published his solution in a paper in the *Acta eruditorum* of April 1689 (see [12] below).

8

On Optical Lines and other matters

Excerpts from a letter to – –.[1]
G. W. Leibniz, *Acta eruditorum*, January 1689[2]

Not long ago, when I was on a rather long trip that my most serene prince ordered me to undertake, and was searching through records in archives and libraries everywhere,[3] a certain friend of mine[4] showed me some monthly issues of the *Leipzig Acta* from which I—having been deprived of new books for a while—learned what was happening in the Republic of Letters. While examining the June issue of 1688, I happened upon a report of the *Mathematical Principles of Nature* by the most illustrious Isaac Newton, which, although it was far from my present [36/37] thoughts, I read eagerly and with great delight.[5] For that man is one of the few who have extended the boundaries of the sciences—a fact that can be demonstrated even just by those series which Nicolaus Mercator of Holstein had obtained by division, but which Newton treated by means of a quite far-reaching invention

[1] From the Latin: '*G. G. L. De lineis opticis et alia; Excerpta ex literis ad – –*'. *Acta eruditorum*, January 1689, 36–8. (ESP 142–5; GM VII 329–31.) Translated by Jeffrey McDonough, Lea Schroeder, and Samuel Levey.

[2] This paper is the first of three papers Leibniz was moved to write after seeing a 12-page report of Newton's *Philosophiæ Naturalis Principia Mathematica* in the June 1688 issue of the *Acta eruditorum*, (303–15), probably written by Christoph Pfautz. See O'Hara (forthcoming, ch. 1). In it Leibniz shows how to construct the catacaustic and diacaustic lines—the 'optical lines' of the title. If rays emanating from a given point are reflected from a mirror of a given shape the catacaustic line is the line describing the shape of a mirror that will reflect those rays back to another given point. The diacaustic is the curve or surface formed by the intersection of refracted light rays. Leibniz shows how to construct the latter using Huygens' notion of an evolute of a curve, this being the envelope of the normals to the curve.

[3] In order to compose the history of the Welfs (or Guelphs) he had been commanded to write by Duke Ernst August, Leibniz had embarked on a fact-finding journey into Central and Southern Germany in November 1687, a journey which eventually took him to Vienna (May 1688–February 1689), and on into Italy (March 1689–March 1690). See Antognazza (2009, 281–319).

[4] This friend was probably the Imperial Librarian Daniel von Nessel, who lent Leibniz books during his stay in Vienna, even bringing them to his room at the 'Steyner Hof' when he was indisposed due to illness (Leibniz 2011, 83; Antognazza 2009, 291).

[5] As Domenico Bertoloni Meli has demonstrated in his book *Equivalence and priority* (1993), Leibniz is being disingenuous here. Although he had indeed written earlier on the topics of these three papers, it is not true that the only access he had to Newton's views was the report in the *Acta*. On the contrary, as is demonstrated by papers in his *Nachlaß* uncovered by Bertoloni Meli, while in Vienna he had access to Newton's *Principia* itself, on which he made notes, and on the basis of which he proceeded to elaborate his own views mathematically, especially on gravitational theory.

Leibniz: Journal Articles on Natural Philosophy. Richard T. W. Arthur, Oxford University Press.
DOI: 10.1093/oso/9780192843531.003.0009

for the extraction of pure and affected roots alike. Quite aside from the transformation of irrational lines into commensurable rational lines (where I call rational those lines whose ordinates can always be obtained from the abscissas in rational numbers), I note here in passing, that, by extending the method of series, I have discovered a procedure [*ratio*] for curves that are given transcendentally, in which there is no place even for [root] extraction. For I assume an arbitrary series and, by treating it according to the rules of the problem, I obtain its coefficients.

Furthermore, I expect brilliant things from Newton's current work, and I see from a report in the *Acta* that it treats not only many utterly new things of very great importance, but also certain other things that I have worked on somewhat. For, aside from the causes of celestial motions, Newton has also attempted to explain catoptric or dioptric lines and the resistance of the medium. Descartes knew about these *Optical Lines* but concealed them, and his commentators did not supply them; for the matter does not fall under common analysis. I understand that they were later discovered by Huygens (though he has not published them yet) and now by Newton. They became known to me as well but, I think, through a different route. Indeed, I have also known general methods for a while, but it was the outstanding discovery, published in the *Acta*, of our Mr Tschirnhaus, who considers whole lines as focal points, that provided the occasion for unearthing very elegant, specific methods. What I derived from this, I will explain by means of an example, from which the rest may be understood.

Let there be a point *A* and a given line *BB* that reflects the rays *AB*; we seek the line *CC* that, in turn, reflects the rays *ABC* to a common point *D*. The solution of a first attempt is the following. If the line *BB* is given, it is clear that the linear confocal, i.e. the line *EE*, of point *A* is given accordingly. Conversely, if two confocals are given—one the linear *EE*, the other the point *D*—it is clear that a line *CC* can be found for which they are focal points, [that is,] the line which was sought. Better constructions appear, however, for $A_1B + {}_1B_1E + \text{arc } {}_1E_2E = A_2B + {}_2B_2E$, and $D_2C + {}_2C_2E + \text{arc } {}_2E_1E = D_1C + {}_1C_1E$, whence the whole $AB + BC + CD$ is always equal to a single constant straight line. If there is a string going around the line *EE* and fastened at point *D*, then [37/38] a pen extending the string will describe the line *CC* through the evolute [*evolutione*] of the curve *EE*. But if the same string is fastened by its other end at point *A*, the pen extending the string will describe the line *BB*.

But, setting aside the curve *EE*, the *easiest construction* turns out to be of the following kind: let an arbitrarily given [line] *AB* be subtracted from a given constant straight line (equal to $AB + BC + CD$); let *BF*, equal to the remainder,

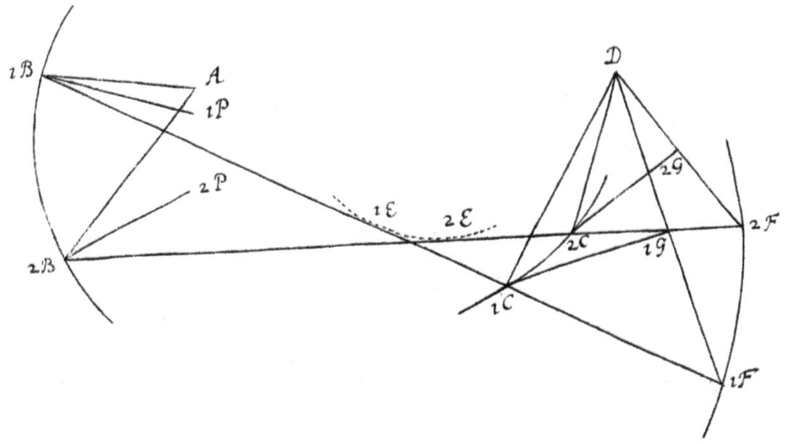

Fig. 8.1

be assumed to have been drawn in such a way that it forms the angle *FBP*, which is equal to *ABP*, with *PB* perpendicular to the curve *BB*, that is, to its tangent at *B*. Let *DF* be joined, and *GC*, drawn perpendicularly from the midpoint *G* of *DF*, will meet *BF* at the point *C* that was to be found; and, indeed, it is evident that *GC* is tangent to the line *CC*. Moreover, let this figure be rotated around the axis *AD*, and the things we have said with respect to lines will also hold with respect to the resulting surfaces [Fig. 8.1].

The same can also be applied to dioptrics. I call a line *EE* that receives the rays *ABC* without reflection and refraction *Acampton*. There are also *Aclastons*,[6] which do not refract the same rays but nevertheless reflect them, and these are the ones that are described by a simple evolute of *EE* (something which Huygens first examined, but to a different end).[7] Such is *FF*, assuming that *CF* (in the extension of *BC*) = *CD*. The same would hold in its own way, if, for the points *A* or *D*, or one of the two points, linear foci were given, or if there were a point infinitely far away, in which case the rays would become parallel.

What I have summarized in a separate paper about the resistance of a medium,[8] I had already worked out for the most part twelve years ago in

[6] Leibniz is here using Latinized Greek terms. The '*acampton*' is ἄκαμπτον, 'unbowed', while the '*aclaston*' is ἄκλαστον, 'unbroken', 'unbreakable'. The former are today called catacaustics, the latter diacaustics, as noted by Hess and Babin (Leibniz 2011, 86, fn. 13). The *evolute* of a curve is the locus of all its centres of curvature. The aforementioned property of the evolute is based on the fact that an evolute is the envelope of the normals to a curve.

[7] Huygens, *Horologium oscillatorium*, 1673, esp. Part III; (1986, 73–104).

[8] Here Leibniz refers to the immediately following article, his *Schediasma de Resistentia Medii, et Motu projectorum gravium in medio resistente*, *Acta eruditorum*, 38–47.

Paris, and I shared some of it with members of the illustrious Royal Academy.[9] Finally, since I also happened to have some thoughts concerning the physical causes of celestial motions, I thought it worthwhile to publish some of them in a separate note. Indeed, I had decided to suppress [*premere*] them until I was able to compare more carefully geometrical laws with the latest phenomena of the astronomers. But (apart from the fact that I am distracted by affairs of an entirely different sort, which hardly allow one to hope for anything like that) Newton's work spurred me to allow these matters to stand as they are so that the sparks of truth emitted from the comparison of the accounts might shine forth all the more, and we might be helped by the acumen of that most ingenious man.

[9] Presumably this alludes to Leibniz's treatise *Du frottement* of Winter 1675, published by H.-J. Hess in *Die unveröffentlichten naturwissenschaftlichen Arbeiten* (1978).

9

A Sketch Concerning the Resistance of a Medium and the motion of heavy bodies projected in a resisting medium[1]

G. W. Leibniz, *Acta eruditorum*, January 1689[2]

When he investigated the rules of the motion of projectiles, Galileo left out the resistance of the medium. Torricelli, and those who followed him, did the same thing, although some of them admit a defect in the doctrine and the errors arising from this in practice. Blondel, indeed, in the book *De Jactu Bomborum*[3] [*On the Projection of Bombs*], thinks this consideration can be safely neglected. But his arguments were insufficient, and the experiments he made were not worth much. Moreover, a geometrical investigation of the matter is more difficult than would have been hoped and readily expected by those people, even though they were very learned, given the wealth of discoveries that had not yet been made at that time, or certainly were not widely enough known. And yet the true laws of projectile motion, and a calculation in agreement with experiments, of such great use in the future for ballistics and artillery, would seem chiefly to depend on this.

Some time ago, I communicated some thoughts on this subject to the celebrated Royal Academy of the Sciences of Paris when I was working among them, where I laid down a way of estimating by parts, and distinguished kinds of resistance.[4] There are, namely, two kinds of resistance of a medium, one absolute and the other respective, which most people usually run

[1] From the Latin: '*G. G. L. Schediasma de Resistentia Medii et Motu projectorum gravium in medio resistente*', *Acta eruditorum*, January 1689, 38–47; Figs. 2, 3, and 4 from Table I, facing p. 37. (ESP 146–156; GM VI 135–43.) Translated by Tzuchien Tho and Richard T. W. Arthur.

[2] As noted in the Introduction, this piece is a further elaboration of Leibniz's previous reflections on the topic in the essay 'Du frottement', which he had prepared for the *Académie des Sciences* in Paris when he was there in 1675, and later discussed with Edme Mariotte during their correspondence in 1683 and 1684. See O'Hara (forthcoming, Introduction).

[3] François Blondel, *L'Art de jetter les bombes* (1683).

[4] The idea of two kinds of resistance, one depending on the viscosity of the medium or friction of surfaces, the other on the density of the medium and the speed imparted to its particles, probably has its

Leibniz: Journal Articles on Natural Philosophy. Richard T. W. Arthur, Oxford University Press.
DOI: 10.1093/oso/9780192843531.003.0010

together. *Absolute resistance* is that which absorbs the same amount of the moving body's force whether it moves with a small or a great velocity, as long as it is moving, and depends on the viscosity [*glutinositate*] of the medium. For it is exactly as if the parts of the medium were connected together by filaments which have to be broken through by the motion of the body. The same holds for the *friction* of rough surfaces on which moving bodies run. For the impediments have to be scraped off, or at least depressed, like elastic hairs that rise up again. The same force must always be expended for depressing the elastic hairs or for breaking the threads, and it does not matter what the acting body's velocity is. *Respective resistance* arises from the density of the medium, and is greater in proportion to the velocity of the moving body. This is because the parts of the medium are agitated by the penetrating body, but moving something depends on a force of penetrating, and this is greater in proportion to how much motion is communicated to the parts of the medium, that is, in proportion to the speed of the penetrating body's motion. And the resistance of a fluid at rest towards a body running into it is equal to the force of the fluid running into the body at rest, which is greater when the motion of the fluid is faster, as we see in the case of bodies moved by the wind and water; indeed, heavy bodies can be supported by a jet of water with sufficient impetus—although here absolute resistance is also mixed in with it, and we should abstract from that when we estimate the respective resistance, as if the medium had no tenacity. Also, there is this big difference between the two kinds of resistance, that absolute resistance has some ratio [39/40] to the surface of the moving body or to the contact; whereas the respective resistance has a ratio to the solidity. In each case a paradox occurs, in that a moving body penetrating into a uniform medium is never in fact reduced to rest by that resistance. As a result of absolute resistance, a body that is once conceived as being moved by a force, and not accelerated by another source, has a certain limit of space or penetration into the medium, so that it always directly approaches that limit but never reaches it. I call this the exclusive maximum penetration, ôr the maximum that it cannot reach. On the other hand, as a result of respective resistance a body that is uniformly accelerated (such as a falling heavy body) has a certain limit of velocity, ôr exclusive maximum velocity, which it always approaches (so that the difference is ultimately insensible), yet never perfectly attains. And this velocity is that very velocity by which the moving fluid

origin in Descartes's account in his *Principles* of a body's motion through the celestial vortex. (See Bertoloni Meli 2006, 301). Only the second (*respective*) resistance would depend on the speed of the body.

(for instance, the jet of water) could support the heavy body without its beginning to fall. We will expound the primary laws of motion for each insofar as brevity may allow, since for the whole thing to be treated distinctly would be a matter for an entire treatise.

On Absolute Resistance

Article 1

If the motion of a body is uniform in itself and retarded by the medium equally with respect to the spaces.

(1) The decrements in the forces are proportional to the increments of the spaces (which is the *hypothesis* of the present case)

(2) The velocities are proportional to the spaces: the past ones to the spaces that have been traversed, the others to the spaces yet to be traversed. Let us suppose the increments of space are equal, then the decrements in the forces will be equal (by proposition 1). Now if the decrements of force of the same moving body are equal, so are the decrements of velocity (for the forces are as the squares of the velocities, while if the existing squares are equal, so are their roots), and so the elements of the velocities that are lost are as the elements of the spaces traversed, and the elements of the remaining velocities are proportional to the spaces to be traversed. Therefore, the velocities are proportional to the spaces. Thus, if in *Fig. 2* [Fig. 9.1] we let the initial velocity be *AE*, and the entire space traversed in the medium be the straight line *AB*, the part of it

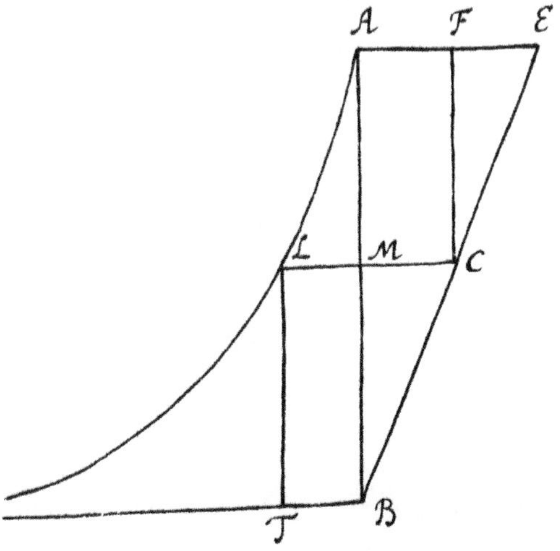

Fig. 9.1

already passed through will be *AM*, that still to be passed through *MB*, the residual speed *MC* (or *AF*), the part of the velocity that is lost *FE*, *ECB* will be a straight line.

(3) If the remaining spaces (*MB* or *LT*) are as the [natural] numbers [40/41], the times spent will be as their logarithms. For if the elements of space are in geometric progression, the residual spaces will be in the same geometric progression, and therefore (by 2) the remaining velocities too. Therefore, the increments of time are equal, so that the times themselves are in arithmetical progression.

(4) The moving body *M* never completes the entire space (*AB*) to be traversed, even if it always approaches the limit *B*. For it is clear that *BT* is the asymptote of the logarithmic line *AL*, that is, the logarithm of *AB* here is of 0, and the logarithm of 0 is infinity. And yet in practice the motion is finally less than can be sensed, so that the distance from *B* is too. Besides, there does not exist anywhere a medium that is perfectly uniform.

(5) If a body moves with a motion composed of an equable one and one equally retarded by the medium with respect to space, that is, if the moving body (*M*) is carried along a rigid straight ruler (*AB*) according to the present hypothesis (while in reality it happens in this way enough because of friction, if the globe moves along the ruler in a horizontal straight line), while meanwhile the ruler (*AB*) itself, moving uniformly while remaining parallel to itself, meets the straight line (*BT*) with one endpoint *B*, it will describe the logarithmic curve (*AL*). For *generally*, if the moving body is carried by a motion composed of a uniform motion and another law, it will describe a line expressing the relation between the times and spaces of the said law by its ordinates and abscissas, which is a *memorable theorem*. Hence we also have a physical way of *constructing logarithms*, which common geometry cannot construct exactly.

Article 2

If the motion is accelerated by gravity, and retarded by the medium equally with respect to the places.

(1) The *first hypothesis* in this case is the same as the unique preceding hypothesis, namely that the decrements in the forces (that is, in this case, of the velocities) brought about by the absolute resistance, are proportional to the increments of the spaces.

(2) The additions of velocities due to gravity are proportional to the increments of the time, and this is the *second hypothesis* about the nature of the motion of heavy bodies.

(3) Let there be straight lines proportional to the assumed times, so that if from each of them an equal straight line is subtracted corresponding to the space traversed by the moving point, the straight line remaining will be proportional to the velocity [41/42] acquired. For the velocities impressed are (by 2) proportional to the times spent traversing the spaces (by 1 here, in the manner proposed in 2 of the preceding article), and therefore proportional to the differences of the acquired remainder.

(4) If the complements of the velocities acquired for the maximum <speed> are as the natural numbers, the times spent will be as their logarithms. Indeed, *retaining the previous figure* 2, let *AB* be the maximum velocity (which, it will soon be clear, is a maximum exclusively), *AM* the acquired velocity, *BM* or *TL* the velocity still to be acquired, that is, the complements of the acquired velocity, *BT* and *ML* the time elapsed; from propositions 1 and 2 it may be learned that when the time increments *BT* are assumed to be equal, the increments of velocity *AM* are proportional to *BM*. Therefore if the times *BT* are in a logarithmic progression, then the velocities *BM* are in an arithmetical one.

(5) Hence it is clear (as was proposed in (4) of the preceding Article) that the maximum velocity *AB* is never attained, that is, that it is a maximum exclusively.

Article 3

If a heavy body is projected in a medium having absolute resistance.

That is, supposing it to be borne by a motion compounded from the motions of the two preceding articles. *In Figure 3* [Fig. 9.2], suppose a heavy body in position *A*, endeavouring to descend along *AG* or along a parallel, is projected from *A* in the direction *AMB* and at some angle *MAG*, and that it describes the line *AP*; let *AB* be the exclusive maximum path of the first article, and let the parallelograms *MAGP* and *BAGK* be completed.

(1) The straight line *BK* perpendicular to the horizontal, where *B* is the limit of penetration (or the point to which the body moving in itself uniformly in a uniform medium with absolute resistance, progressing along the straight line *AM*, is unable to penetrate) is the asymptote of the line of projection, that is, of the two lines, namely the straight line *BK* and the curve AP: however far these are continued, even though they always approach each other, they never touch each other. This is because the moving body tends towards *B* along *AB*, and in the same way along the parallels, with a composite motion, as if it were carried

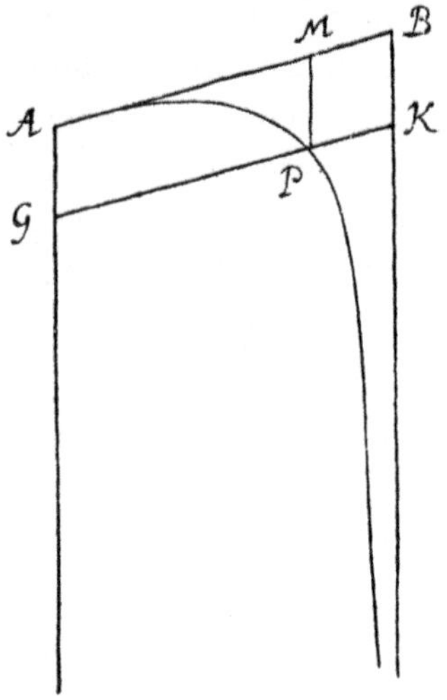

Fig. 9.2

without gravity according to the laws of the first article alone, and therefore it never reaches *B* or any equivalent point on the straight line *BK* when produced.

(2) The line of projection is not one of the conic sections, nor is it by any means a parabola, a circle, or ellipsis, since these lack asymptotes. It is not a hyperbola, however, for here one cannot take still other lines as asymptotes through some point [42/43] which is taken on a straight line *BK* unbounded at both ends, as one can in the hyperbola.

(3) Let there be a certain *simple line* (such as a parabolic or hyperbolic one), such that if its abscissas are proportional to the remaining spaces (*BM*) to the prescribed limit of penetration (*AB*) of the projection, the ordinates are proportional to the velocities thus far lost for attaining the prescribed limit of the velocity of descent. Here I understand a *simple curve* to be that whose ordinates are in any multiple or submultiple ratio of the abscissas. Thus, the sense is that the velocities lost in the descent up to this point are in the ratio of the spaces that are still short of the limit of penetration, multiplied by some constant number. This is demonstrated from the fact that both can be understood to be in geometric progression. If the elapsed times are in arithmetical progression by art. 1, prop. 3 and art. 2, prop. 4, and in both the logarithm of the maximum number is 0, then the logarithm of the minimum number is infinite, by art. 1, prop. 4 and art. 2, prop 5. When the number multiplying the ratio is rational, there arises a parabolic or hyperbolic curve of common geometry. Moreover, this number can be discovered by certain experiments.

(4) One can discover the *line of projection AP*, that is, the relation between the coordinates *AG* of the space descended and the space *AM* of a per se uniform progression. For by art. 2, prop. 3 a simple relation is given between the time elapsed, the space *AG* traversed in the descent, and the velocity of descent acquired at G. In this relation *AM* is substituted for the time, by virtue of the relations given in art. 1, prop 3. The relation between *AG* and *AM* therefore still holds, which, even though it is transcendental, nonetheless presupposes nothing other than logarithms.

Article 4

On the respective resistance of the medium, if a motion in itself uniform is retarded by a uniform medium in proportion to velocity.

It is considered only with respect to the density of the medium, without taking account of its tenacity.

(1) The diminutions of the velocities are in a ratio compounded from the present velocities and the increments of space. This is the *hypothesis* of the present case. [43/44]

(2) If the residual velocities (for instance, *MB* or *LT* in *fig.* 2) are in arithmetical progression, the spaces traversed (*BT* or *ML*) are in a logarithmic one. This is demonstrated in the same way as in art. 1, prop. 3, if instead of the supposed spaces you put velocities, and instead of times, spaces.

(3) If the times elapsed, increased by a certain constant quantity, are in arithmetical progression, then the spaces traversed are in a logarithmic one. For with the existing elements of space equal, the existing elements of time are inversely as the velocities, that is, they increase in a geometrical progression (by the preceding), therefore (by a logarithmic quadrature) the times that are increased by a constant quantity are in geometrical progression.

(4) Hence also the times are increased by a constant quantity in inverse proportion to the residual velocities, as is clear from the preceding consideration. Now that constant quantity is the finite time in which an infinite distance is traversed, if the first velocity increases in the same proportion as it is now diminished by the resistance of the medium. And this quantity can be found by means of two experiments, by comparing together the spaces and the times, and even in a single experiment, where time and velocity are considered.

Article 5

If a motion accelerated by gravity is retarded by a uniform medium in proportion to velocity.

(1) *Hypothesis 1* in this case is the same as the first hypothesis of the preceding article.

(2) And *hypothesis 2* is the same as hyp. 2 of the second article.

(3) The resistance to a new impression made by gravity in the same element of time (that is, the diminution of the velocity of approach) is in the ratio of the square of the excess of the maximum velocity over the acquired velocity to the square of the maximum velocity. For by prop. 1 of this article it follows that the resistances are in the compound ratio of the elements of time and the squares of the velocities; but the new impressions are as the elements of time by prop. 2, and in the case of the maximum the diminution and increase of velocity are equal. From which the proposition is easily concluded.

(4) If the ratios between the sum and difference of the assumed maximum and minimum velocity are in arithmetical progression, then the times in which the assumed velocities are acquired will be logarithmic. For since the increment of [44/45] velocity is the difference between the impression and the resistance, from this (by the preceding) it follows immediately that the impression is to the increment of velocity, as the square of the maximum velocity is to the excess of its square over the square of the present assumed velocity. From which we know by quadratures that the sum of the impressions from the beginning, which is proportional to the time elapsed, is logarithmic, if the number is of the kind we have enunciated in this proposition.

(5) The maximum velocity is so exclusively, that is, it can never be attained, even if it approaches it in an unassignable interval. For when the ratio is one of equality, that is, when the assumed velocity is just beginning or infinitely small, the time (and thus the logarithm) is 0, and therefore when the ratio is infinite, that is, when the assumed velocity is itself a maximum, the logarithm of the ratio is infinite. And so it would require an infinite time for this velocity to be acquired. The maximum speed can be found by two experiments, comparing times and speeds, likewise by prop. 3.

(6) If the acquired velocities (*AV* in *fig. 4* [Fig. 9.3]) are as the sines (of the portions of the arc *HK* of the quadrant of the circle *HKB*) the spaces traversed (*AS*) will be as the logarithms of the complement (*VK*) of the sines, supposing the radius or total sine (*AB*) to be proportional to the maximum velocity. For by hypothesis 2 it follows that the increments of space are in the compound ratio of acquired velocities and impressions of gravity, but the impressions are

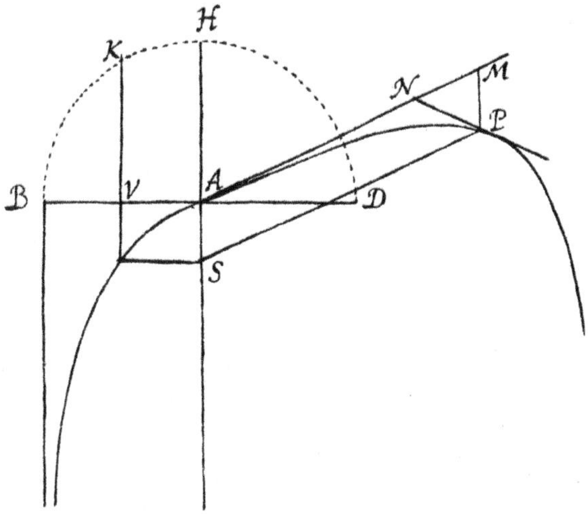

Fig. 9.3

as the increments of velocity, as was stated in the demonstration of prop. 4. Hence it follows that the increments of space are in the compound ratio of the increments of velocity and the velocities taken directly, and in the inverse ratio of the excess of the square of the maximum speed over that of the assumed squares. From which we know by means of quadratures that the proposition follows. It is clear from this that the logarithm of the total sine is 0 (when the velocity is 0), but the logarithm or space of the complement of the vanishing sine (when the velocity is maximal) is infinite, from which in turn it is clear that the maximum velocity can never be attained.

(7) If the spaces traversed (*As* in *fig. 4*) are as the logarithms of the sines (*KV*, of the arc *BK*), then the times elapsed are as the logarithms of the ratios which between the versed sine (*BV*) and its complement (*VD*) have towards the diameter (*BD*), or twice the total sine (*AB*). This is clear from propositions 4 and 6 taken together. [45/46]

Article 6

If a heavy body is projected in a uniform medium having a respective resistance.

Or if it is carried by a motion compounded from the two motions of the preceding articles. Let the projection (*fig. 4*) be along *AM* and its parallels, the descent along *AS* and its parallels, and the angle *MAS* be any angle whatsoever; a position *P* in the compound motion is obtained by completing the parallelogram *MASP*.

(1) The line of projection (or the relation between *AS* and *AM*) can be found. The velocity *AV* of descent to S or to P is given by the space *AS* (by art. 5, prop. 6). From this (by prop. 7) the time elapsed is obtained. From this (by art. 4, prop. 3) the space *AM* or *SP* is given. Therefore, from the given abscissa lines *AS*, the ordinates *SP* are given, and thus the points of the line can be found.

(2) One can find the tangent of the line, that is, the direction of the body moving on it. In *AM* let us assume *MN*, which is to *MP* as the velocity at *M*, found by means of the elapsed time (by art. 4, prop. 4), is to the velocity at *S*, found by means of the same time (art. 5, prop. 4), and when *NP* is joined it touches the curve at the point *P*. And since the velocity of descent at *P* is the same as at *S*, and likewise since the velocity of continuing in the direction of projection at *P* is the same as at M, it is clear what this becomes at point *P*; it is also clear by what force the moving body is carried along the line of projection itself, for the velocity along this line is to the velocity of descent, as *NP* to *MP*.

We could also combine into one the absolute resistance from articles 1, 2, and 3, and the respective resistance from articles 4, 5, and 6, as they certainly do in fact occur together in nature, but we should avoid being long-winded here. Many practical applications may be deduced from these considerations, but now for us it will have been enough to have established geometrical foundations, which is where the greatest difficulties lie. And perhaps we will be seen by careful consideration to have uncovered certain new paths, or at least things that were previously obstructing us. But all this is in agreement with *our analysis of the infinite*, that is, *the calculus of sums and differences* (some elements of which we have given in these *Acta*), insofar as it can be expressed here in common terms.

Having seen some things concerning the use of gunpowder in machines in the *Acta* of Leipzig as well as in the *New Acts* of Rotterdam, I would say in the first place that the most celebrated Thévenot has had some thoughts about this matter concerning hydraulics with which I agree,[5] and from this there also occurred to me some matter for further meditation, [46/47] which would comprise *Περιάνδρου Κλεανθης* [*Cleanthes On Man*].[6] I wanted to add this here in passing.

[5] Melchisédech Thévenot (c. 1620–1692) was a French natural philosopher and man of means, a friend of Christiaan Huygens and mentor of Jan Swammerdam and Niels Stensen. Thévenot wrote on hydrostatics in his essay 'Discours sur l'art de la navigation' contained in *Recueil de voyages de Mr Thévenot* (Thévenot 1681). Leibniz had extensive correspondence with Thévenot until the latter's death in 1692, especially while he was composing his own *Protogaea* (Leibniz 1749), concerning geology and the natural-historical development of the Earth from its origins. See O'Hara (forthcoming, Introduction).

[6] Cleanthes of Assos (330–230 BCE) was the successor to Zeno of Citium and second head of the Stoic school in Athens. He was known for his teaching that matter is not dead and inert, but has a characteristic tension (τονος). This perhaps explains the connection with the topic of this chapter, the resistance of the medium, but not why Leibniz thought it worth mentioning that he was thinking of writing such a meditation.

10

An Essay on the Causes of the Celestial Motions[1]

Acta eruditorum, February 1689[2]

It is well known that the ancients, especially those who followed the opinions of Aristotle and Ptolemy, did not recognize the majesty of nature, which has at last shone forth in our century and the preceding one, ever since Copernicus, having recalled from obscurity the most beautiful hypothesis of the Pythagoreans—which they seem perhaps to have come upon by guesswork rather than correctly established—showed it to satisfy the phenomena with the utmost simplicity. Moreover, Tycho, having followed Copernicus in the main points of his system (except in the transposing of the Sun and the Earth), turned his mind to more accurate observations than was customary, and removed from the heavens the wholly unseemly apparatus of solid orbs. But even though he did not sufficiently perceive the fruits of his Herculean labours—partly because he was hindered by certain prejudices, partly because he was prevented from doing so by his death—divine providence brought it about that his observations and efforts came into the hands of that incomparable gentleman Johannes Kepler, whom the Fates had destined to be the first among mortals to make public

[1] From the Latin, *Tentamen de Motuum Coelestium causis, autore G.G.L. Acta Eruditorum*, February 1689, 82–96. (ESP 157–172; GM VI 144–61 (published version); 161–87 (unpublished second draft)). Translated by Richard T. W. Arthur.

[2] As described in the Introduction, this essay was occasioned by Leibniz's reading of Pfautz's 1688 report of Newton's *Principia*; he subsequently had Newton's book in his possession for long enough to make some notes on it, before writing several preparatory drafts to this essay. See Bertoloni Meli (1993) for a full account. It comprises Leibniz's theory of celestial motions, including a derivation of the mathematical formula for the solicitation away from the centre of a body moving under the action of a central force in accordance with Kepler's first two laws. Leibniz also includes a paragraph describing his interpretation of the differences (or differentials) of quantities, which he ever after refers to as his *Lemmas on Incomparables*. He notes that although this treatment is easily generalizable to parabolic and hyperbolic orbits, he has not given a derivation of Kepler's third law, nor a hypothesis to explain the inverse square law.

Leibniz: Journal Articles on Natural Philosophy. Richard T. W. Arthur, Oxford University Press.
DOI: 10.1093/oso/9780192843531.003.0011

the arrangement of heaven, the order of things, and the laws of the gods.[3]

He therefore discovered that any primary planet describes an elliptical orbit in which the Sun occupies one of its foci, with the law of motion that, when the radii are drawn from the Sun to the planet, the areas cut out by them are always proportional to the times. The same man determined that the various planets in the same system have periodic times that are in the sesquialterate ratio of their mean distances from the Sun, and would surely have celebrated his triumph wonderfully had he known (as Cassini above all observed) that even the satellites of Jupiter and Saturn follow the same laws with respect to their planets as the latter do toward the Sun. But he could not yet provide the causes of so many and such unvarying truths, both because his mind was preoccupied with Intelligences or unexplained radiations of sympathies, and also because [82/83] in his time higher geometry and the science of motions were not yet as far advanced as they are now. Yet he also opened the way for the investigation of causes [*rationes*]. For to him we owe the first indication of the true cause of gravity, and of the law of motion on which gravity depends, which is that rotating bodies endeavour to recede from the centre along the tangent; so that, if straws or husks are floating in water, then when the vessel is rotated the water forms in a vortex, and being denser than the straws and so being driven away from the middle more strongly than them, it pushes them towards the centre. This is just what Kepler himself clearly explains in two or more places in his *Epitome Astronomiae*, although, still beset by doubts and ignorant of his own resources, he was not sufficiently aware of what would follow from this in physics and especially in astronomy. But later *Descartes* made brilliant use of these considerations, even if, as was his custom, he pretended to be their author. Moreover, I often marvel that Descartes did not even attempt to provide reasons for the celestial laws discovered by Kepler, so far as we know, whether because he could not sufficiently reconcile them with his own opinions, or because he was ignorant of how happy a discovery this was, and did not think that the laws were so closely observed by nature.

Furthermore, it seems scarcely physical, and indeed unworthy of the admirable machine-work of God, to assign individual *Intelligences* to the stars to direct their courses, as if God were lacking the reasons for bringing about the same things by laws governing bodies; and in truth *solid orbs* were refuted

[3] Here Leibniz quotes a famous line from the poet Claudian's eulogy on Archimedes' Sphere (a clockwork device modelling the motions of the heavens). A poetic translation of the couplet in question, '*Jura poli rerumque fidem legesque Deorum/ Ecc Syracusus transtulit arte senex*', is 'An aged Syracusan, by his skill / Arranges poles, laws, harmony at will'.

some time ago, while *sympathies*, magnetisms, and other abstruse qualities of that kind are either not understood, or when they are, judged to be the effects that would appear from corporeal impressions. These things being so, I myself judge that there is no other alternative than that the cause of the celestial motions should originate from the motions of the aether, or, to speak like an astronomer, from *orbs* that are *deferent*, but *fluid*. This opinion is very ancient, although neglected: for Leucippus expressed it prior to Epicurus, insofar as he used the name δίνης (*vortex*) in fashioning his system, and we have heard how Kepler foreshadowed gravity by the motion of water moved in a vortex. And from the *Voyages* of Monconys[4] we learn that Torricelli (and, as I suspect, also Galileo, whose disciple he was) was already of the opinion that the whole aether with the planets is driven round by the motion of the Sun about its centre, just as water is driven round by a stick rotated in the middle of a vessel at rest; and just like chaff or straws floating in the water, the stars that are nearer the middle circulate faster.

But these more general considerations come to mind [83/84] without much difficulty. It is our intention, however, to explain the laws of motion themselves more distinctly, which will prove to be a matter for a much more profound investigation. And since we have thrown some light on this kind of thing, and the inquiry seems to have gone extremely well and quite naturally, this has raised my hope that we have come close to the true causes of celestial motions.

(1) In order to tackle the matter itself, it can be demonstrated first that according to the laws of nature *all bodies that describe a curved line in a fluid are set in motion by the fluid itself*. For given that all bodies describing a curve endeavour to recede along a straight line (by the nature of motion), there therefore needs to be something to constrain them. But there is nothing contiguous to them except the fluid (by hypothesis), and no endeavour is constrained except by something contiguous and moving (by the nature of body); therefore it is necessary that the fluid itself be in motion.

(2) Hence it follows that *the planets are moved by their own aether*, that is, that they have fluid orbs that are deferent or moving. For by universal consent they describe curved lines, and it is impossible to explain the phenomena by supposing rectilinear motions alone. So (by the preceding) they are moved by an ambient fluid. The same thing can be demonstrated in another way from the fact that the motion of a planet is not equable, that is, it does not describe

[4] Balthasar de Monconys (1611–1665) was a French Royal Counsellor who travelled widely, and recorded his travels in his *Journal des Voyages* (Monconys 1665).

equal spaces in equal times. From this, too, it follows that it must be driven by the motion of an ambient fluid.

(3) I call a *circulation* a *harmonic* one if the velocities of circulation that are in some body are inversely proportional to the radii ôr distances from the centre of circulation, or (what comes to the same thing) if these velocities of circulation around the centre decrease in proportion to how the distances from the centre increase, or very briefly, if the velocities of circulation increase in proportion to their proximity to the centre. For in this way if the radii ôr distances increase equably ôr arithmetically, the velocities will decrease in a harmonic progression. And so harmonic circulation can be found to occur not only in circular arcs, but also in any other curve one might describe. Suppose (see Table II, Figure 1 [Fig. 10.1])[5] the moving body *M* is carried in a certain

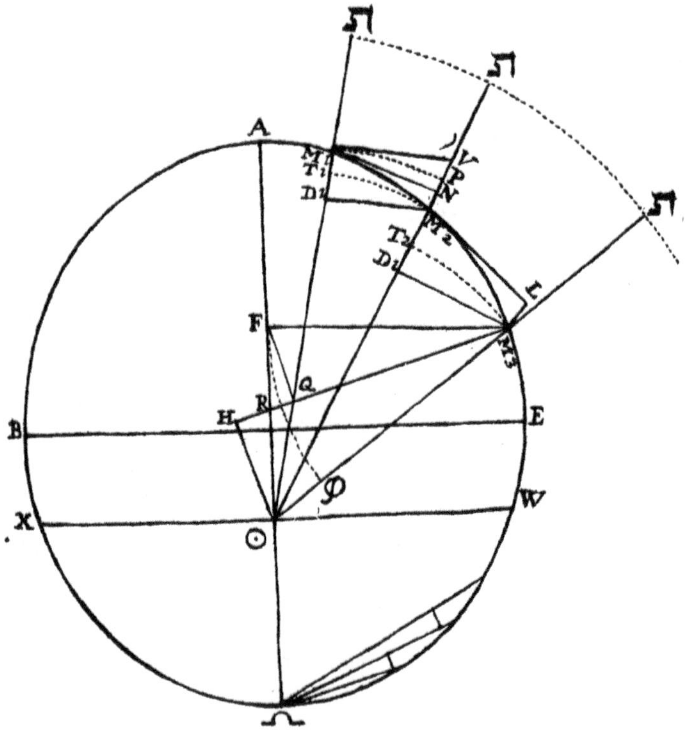

Fig. 10.1

[5] See chapter 24 below for Leibniz's corrections to certain shortcomings of this figure.

curve ${}_3M_2M_1M$ (or ${}_1M_2M_3M$)[6] and in equal elements of time describes the elements of the curve ${}_3M_2M$ and ${}_2M_1M$; then its motion can be understood to be composed of a circular one around some centre, such as $\odot$ [the Sun] (for instance, ${}_3M_2T$, ${}_2M_1T$) and a rectilinear one such as ${}_2T_2M$, ${}_1T_1M$ (with $\odot_2T$ taken equal to $\odot_3M$, and [84/85] $\odot_1T$ to $\odot_2M$). This kind of motion can also be understood when a ruler or indefinite rigid straight line $\odot n$ moves around the centre $\odot$, and meanwhile the moving body is moved along the straight line $\odot n$. Moreover it does not matter what the rectilinear motion is by which it approaches the centre or recedes from it (I call this *paracentric motion*), provided that the circulation of the moving body M, e.g. ${}_3M_2T$, is to another circulation of it, e.g. ${}_2M_1T$, as $\odot_1M$ to $\odot_2M$, that is, if the circulations made in equal intervals of time are inversely as the radii. For since these arcs of elementary circulations are in the composite ratio of times and velocities, whereas their elementary times are assumed to be equal,[7] their velocities will thus also be inversely as the radii, and so the circulation is called harmonic.

(4) If the moving body is carried in a harmonic circulation (whatever its paracentric motion may be) then the areas cut out between the radii drawn from the centre of circulation to the moving body will be proportional to the times taken, and vice versa. For since the elementary circular arcs, such as ${}_1T_2M$, ${}_2T_3M$, are incomparably small in relation to the radii $\odot_2M$, $\odot_3M$, the differences between the arcs and their sines (such as between ${}_1T_2M$ and ${}_1D_2M$) will be straight lines incomparable with the differences themselves, and therefore (by our analysis of the infinite) they are taken to be null, and the arcs and sines to be coincident. Therefore ${}_1D_2M$ is to ${}_2D_3M$ as $\odot_2M$ is to $\odot_1M$, ôr $\odot_1M$ times ${}_1D_2M$ equals $\odot_2M$ times ${}_2D_3M$. Therefore their halves are also equal, namely the triangles ${}_1M_2M\odot$ and ${}_2M_3M\odot$. Since these triangles are elements of the area $A\odot MA$, and the elements of time are taken equal by hypothesis, the elements of area are also equal, and vice versa, and therefore the areas $A\odot MA$ are proportional to the times in which the arc AM is traversed.

(5) In demonstrating these things I have assumed *incomparably small quantities*, for example, the difference between two common[8] quantities

[6] Leibniz characteristically wrote the index as a prefix—thus ${}_1M$—where we would make it a suffix—thus M_1. Confusingly, however, on the diagram his points ${}_1M$, ${}_2M$, etc. are rendered as M_1, M_2 etc. He does not comment on this inconsistency in his 'Excerpt from a Letter of G.G.L.' when he lists corrections to the *Tentamen* and its accompanying figure, seemingly regarding them as interchangeable.

[7] Here we have given a very literal translation; less literally, 'since these arcs of the elements of circulation are as the products of times and velocities, while the elements of time are assumed to be equal, . . .'.

[8] By 'common' Leibniz means finite, assignable as opposed to unassignable.

incomparable with the quantities themselves. Now if I am not mistaken, such things can be lucidly expounded as follows. Thus if someone does not wish to employ *infinitely small* quantities, one can assume them as small as one judges sufficient for them to become incomparable, and to produce an error of no importance, indeed, an error smaller than any given. Just as the Earth is taken as a point, ôr the diameter of the Earth is taken as an infinitely small line with respect to the heaven, in the same way it can be demonstrated that if the sides of an angle have a base that is incomparably smaller than them, [85/86] the angle comprised will be incomparably smaller than a right angle, and the difference between the sides will be incomparable with the differences themselves. Likewise, the difference of the whole sine, of the sine of the complement and of the secant, will be incomparable with the differences, and likewise the difference of the sine,[9] the chord, the arc, and the tangent. Therefore, since these quantities are themselves infinitely small, their differences will be *infinitely many times infinitely small*, and the versed sine will also be infinitely many times infinitely small, and so incomparable with a right angle. And there are infinitely many degrees of the infinite, and just as many degrees of the infinitely small. And one can use common triangles that are similar to those *unassignable* ones, which are of the greatest use in finding tangents, and maxima and minima, and for working out the curvature of lines; and likewise in almost every application of geometry to nature. For, if motion is expressed by a common line that a moving body traverses in a given time, impetus ôr velocity is expressed by an infinitely small line, and the element of velocity, such as the solicitation of gravity or centrifugal endeavour, is expressed by an infinitely many times infinitely small one. And I reckoned that these things ought to be noted down here as *lemmas* for *our Method of incomparable quantities* and *analysis of infinites*, and as it were *Elements* of this new doctrine.

(6) From these things it is already a consequence that *the* primary *planets move in a harmonic circulation* around the Sun as centre, and satellites around their primary planet. For with the radii drawn from the centre of circulation they describe areas proportional to the times (as per observations). Therefore, with the elements of the times supposed equal, the triangle ${}_1M_2M\odot$ is equal to the triangle ${}_2M_3M\odot$, and therefore $\odot_1M$ to $\odot_2M$ is as ${}_2D_3M$ to ${}_1D_2M$, that is, the circulation is harmonic.

(7) It is also agreeable that the aether ôr *fluid orb of each planet moves by a harmonic circulation*. For it was shown above that no body in a fluid moves

[9] Here 'the sine' should be deleted, as Leibniz notes in 1706.

spontaneously in a curved path, so there will also be a circulation in the aether, and it is reasonable to believe that this will be in agreement with the circulation of the planet, so that the circulation of the aether of each planet is also harmonic. That is, the fluid orb of each planet is divided in thought into innumerable concentric circular orbs of negligible thickness, each of which will have its own circulation that is faster in proportion to how close it is to the Sun. But a more accurate account of this motion in the aether will be given in due course.

(8) Thus, we suppose that a planet is moved with a double motion, one composed [86/87] of a harmonic circulation of its fluid deferent orb, and a paracentric motion as if possessed of a certain gravity ôr attraction, that is, an impulse towards the Sun or primary planet. Now the circulation of the aether makes the planet circulate harmonically, not by its own motion, but by floating calmly, as it were, in the deferent fluid whose motion it follows. Hence it does not retain the faster impetus of circulating it had in its inferior or closer orb, but becoming slower as it traverses the superior orbs (by resisting a velocity greater than its own), it continuously loses impetus and insensibly accommodates itself to the orb it approaches. Conversely, when it tends from the superior orbs towards the inferior ones, it receives their impetus. And this happens all the more easily because, once the motion of the planet agrees with the motion of the orb it is in, afterwards it hardly differs from the closest ones.

(9) Having explained the harmonic circulation, we must come to *the paracentric motion of the planets arising from the outward impression of the circulation and the solar attraction* in combination with each other. Now one may call it an attraction, although it is really an impulse, for indeed the Sun can be conceived in a certain sense as a magnet; the magnetic actions themselves, however, are undoubtedly derived from the impulses of fluids. Whence we shall also call it the *solicitation of gravity*, conceiving the planet as a heavy thing tending to the centre, namely the Sun. The type of orbit, however, depends on the particular law of attraction. Let us see then what law of attraction makes an elliptical path. In order to achieve this, we need to enter the sanctuary of geometry for a while.

(10) Since every moving body that describes a curved line endeavours to recede along the tangent, one may call this the *outward endeavour*, as in the motion of a sling, for which there is required an equal force which constrains the moving body so that it does not fly away. *This endeavour may be measured by the unassignable perpendicular distance of a preceding point in the tangent from the following point.* And when the line is circular, this *force* collected from

the repeated endeavours[10] was called *centrifugal* by the most celebrated Huygens, who first treated it in geometry. Now, every outward endeavour is infinitely small with respect to the velocity ôr impetus acquired from the endeavour when repeated for some time, just like the solicitation of gravity, whose nature is homogenous with it; by which it is also confirmed that each of them has the same cause. Nor, therefore, is it any wonder that percussion, as Galileo maintained, is infinite in comparison with bare gravity, or as I would say, to simple [87/88] endeavour, the force of which I am accustomed to call *dead*, which, by acting while receiving impetus from repeated impressions, is finally rendered *live*.

(11) *Centrifugal endeavour,*[11] ôr the outward endeavour of circulation, *can be expressed by PN, the versed sine of the angle of circulation* $\odot_1 MN$ (or by ${}_1D_1T$, which turns out as the same because the difference between the radii is unassignable); for the versed sine is equal to the perpendicular from one endpoint of the arc of the circle onto the tangent from the other drawn [radius], by which we represented the outward endeavour in the preceding paragraph. (The centrifugal endeavour can also be represented by *PV*, the difference between the radius and the secant of the same angle, since the difference between this difference and the versed sine is *infinitely, infinitely, infinitely small,* and therefore completely negligible with respect to the radius.) Hence, moreover, the versed sine is as the square of the chord, that is, the square of an unassignable arc ôr velocity, it follows that *the centrifugal endeavours of moving bodies describing equal circles with an equable motion are as the squares of the velocities, and those of bodies describing unequal circles are as products of the squares of the velocities with the reciprocals of the radii.*

(12) *The centrifugal endeavours of a body circulating harmonically are inversely as the cubes of the radii.* For (by the preceding) they are inversely as the radii, and directly as the squares of the velocities, that is (since the velocities of the harmonic circulation are inversely as the radii), inversely as the squares of the radii; but from the simple inverse and the inverse square, we get the inverse cube ratio. For the calculation, let θa be a constant area always

[10] In his 1706 letter Leibniz advises that the phrase 'collected from the repeated endeavours' should be omitted.

[11] As Leibniz explains in his 1706 letter, 'centrifugal endeavour' should more appropriately be the endeavour of the motion along the arc, which is twice that of the endeavour at the tangent (strictly speaking), which is the meaning he adopts in this essay. So everywhere in paragraphs 11, 12, 15, 21, 27, and 30 where he has written 'twice the centrifugal endeavour' it should read simply 'centrifugal endeavour', and everywhere he has written simply 'centrifugal endeavour' it should read 'half the centrifugal endeavour'. He takes motion along the arc to be equal to that along the chord (since the difference is unassignable), so that the direction of the continuation of this motion is along the prolonged chord.

equal to twice the elementary triangle ${}_2M_3M\odot$, ôr to the rectangle ${}_2D_3M$ times the radius $\odot_2M$ ôr r. Then ${}_2D_3M$ will be $\theta a{:}r$ ôr θa divided by r. Now the centrifugal endeavour ${}_2D_2T$ is equal to the square of ${}_2D_3M$ divided by twice $\odot_3M$, and therefore equals $\theta\theta aa{:}2r^3$.

(13) *If the paracentric motion* (receding from a centre Ω or approaching it) *is uniform,* and the circulation harmonic, *the line of motion ΩMG will be a spiral* beginning from the centre Ω, having the property that the segments *ΩMGΩ are proportional to the radii,* that is, in this case, *to the chords ΩG* drawn from the centre. For as the areas, that is, the segments, are proportional to the times, so (because of the equable recession) are the radii. There are many other notable properties of this spiral, and its construction is not difficult. Indeed, in harmonic circulation *there is a general method* for constructing the lines, at least if their quadratures are supposed, if the times, or the velocities of the paracentric motion, or at least [88/89] the elements of impetus ôr solicitations of gravity, are given by the radii.

(14) *The paracentric solicitation, whether of gravity or levity,* is expressed by the straight line ${}_3ML$ made from the point ${}_3M$ of the curve onto the tangent ${}_2ML$ (produced to L) from the unassignably distant preceding point ${}_2M$ and parallel to the preceding radius $\odot_2M$ (drawn from the centre $\odot$ to the preceding point ${}_2M$).

(15) *In every harmonic circulation the element of paracentric impetus* (that is, the increment or decrement of the velocity of descent towards the centre, or of ascent from the centre) *is the difference or sum of the paracentric solicitation* (that is, of the impression made by gravity or levity or a similar cause) *and twice the centrifugal endeavour* (arising from the harmonic circulation itself); the sum, that is, if levity is present, the difference if gravity is; where, when the solicitation of gravity prevails the velocity of descent increases or that of ascent decreases, but when twice the centrifugal endeavour prevails, it is the other way around. From ${}_1M$ and ${}_3M$ let ${}_1MN$ and ${}_3MD$ be normals onto $\odot_2M$. Then, since the triangles ${}_1M_2M\odot$ and ${}_2M_3M\odot$ have been shown to be equal because of the harmonic circulation, their heights will also be equal (because of their common base $\odot_2M$). Now, taking ${}_2MG$ and L_3M equal, let ${}_3MG$ be joined parallel to ${}_2ML$. Then the triangles ${}_1MN_2M$ and ${}_3M_2DG$ will be congruent, and ${}_1M_2M$ will equal G_3M, and N_2M will equal G_2D. Next, in the straight line $\odot_2M$ (produced, if need be, as is always understood) let $\odot P$ be taken equal to $\odot_1M$, and $\odot_2T$ equal to $\odot_3M$. Then P_2M will be the difference between the radii $\odot_1M$, and $\odot_2M$, and ${}_2T_2M$ the difference between the radii $\odot_2M$, and $\odot_3M$. Now P_2M equals (N_2M ôr) $G_2D + NP$, and ${}_2T_2M$ equals ${}_2MG + G_2D - {}_2D_2T$, so $P_2M - {}_2T_2M$ (the difference of the differences) will be $NP + {}_2D_2T - {}_2MG$, that is

(since NP and ${}_2D_2T$, the versed sines of two angles and radii whose differences are incomparable, coincide) equals twice ${}_2D_2T - {}_2MG$. Now the difference of the radii expresses the paracentric velocity, the difference of the differences expresses the element of paracentric velocity. Moreover, ${}_2D_2T$ ôr NP is the centrifugal endeavour of the circulation, being the versed sine (by 11) and ${}_2MG$ or ${}_3ML$ is the solicitation of gravity (by the preceding). Thus the element of paracentric velocity equals the difference between twice the centrifugal endeavour NP ôr ${}_2D_2T$ and the simple solicitation of gravity G_2M, or (as is concluded in the same [89/90] way) the sum of twice the centrifugal endeavour and the simple solicitation of levity.

(16) When the increments or decrements of the velocity of ascent or descent are given, so also are the solicitation of gravity or levity, and vice versa. This is clear from the preceding, for I hold that the endeavour is always given, since it is inversely as the cube of the radii (by 12).

(17) In equal elements of time the increments of the angles of harmonic circulation are inversely as the squares of the radii. For the circulations are in the compound ratio of the angles and of the radii, and the elementary circulations, since they are harmonic, are inversely as the radii, therefore the elementary angles are inversely as the squares of the radii. Such are more or less the apparent daily motions as seen from the Sun (for the days are here sufficiently minute parts of time, especially for the more distant planets), which will be approximately in the ratio of the inverse squares of the distances, so that at a double distance only a quarter of the angle would be covered in the same element of time, at a triple distance, a ninth.

(18) If an ellipse is described by the harmonic circulation of a body around a focus as centre of circulation, the circulation ${}_2T_3M$ or ${}_2D_3M$ (for these differ by an incomparable quantity), the paracentric velocity ${}_2D_2M$, and the velocity of the moving body (composed from these) in the elliptical orbit itself, namely ${}_2M_3M$, are respectively as these three quantities: the transverse axis BE; the mean proportional between the difference and the sum of the distance $F\odot$ of the foci from one another, and the differences $\odot\varphi$ of the distances of the point ${}_3M$ of the orbit from the foci; and lastly, twice the mean proportional between $\odot_3M$ and F_3M, the distances of the same point from the two foci.[12] These same things are true in the hyperbola in its own way. In the parabola, with the

[12] As Domenico Bertoloni Meli explains (1993, 136, fn.17), this means that the paracentric velocity ${}_2D_2M = \sqrt{\{(F\odot + \odot\varphi) \cdot (F\odot - \odot\varphi)\}}$, while the orbital velocity ${}_2M_3M = 2\{\odot_3M \cdot F_3M\}$. Leibniz has set the arc ${}_2T_3M$ equal to the chord ${}_2D_3M$, and the straight line ${}_2T_2M$ equal to ${}_2D_2M$, since in each case those quantities differ by a quantity that is incomparable with them; that is, infinitely small in comparison with them, and thus null.

quantities that are infinite there vanishing, the circulation, the paracentric velocity, and the velocity that is composed of them which is in the orbit itself, become proportional to, respectively: the latus rectum; the mean proportional between the latus rectum and the excess of the radius over the minimum of all the radii (which is a quarter of the latus rectum); and finally, twice the mean proportional between the radius and the latus rectum. The truth of these claims can be derived from the common elements of conic sections if one supposes that the straight line ${}_3M$[13] perpendicular to the curve (or its tangent) at ${}_3M$ meets the axis $A\Omega$ in R, and that the normals FQ and $\odot H$ are drawn to it from the focus; then it is clear that ${}_3MH$, $H\odot$, and [90/91] $\odot_3M$[14] are proportional to ${}_2M_2D$, ${}_2D_3M$, and ${}_3M_2M$, that is, to the paracentric velocity, the circulation, and the velocity in the orbit itself. Therefore, it suffices to show that the sides of the triangle ${}_3MH\odot$ are to each other as we have stated. This will be done more easily by considering that the triangles ${}_3MQF$ and ${}_3MH\odot$ are similar, and moreover that F_3M is to $\odot_3M$ as FR to $\odot R$, from which what was proposed may be inferred by the common analysis. From this it follows that, allowing the foci to be interchanged so that one becomes the centre of the harmonic circulation and attraction instead of the other, the ratio between the circulation and the paracentric velocity will remain the same as before.

(19) If a moving body that has gravity, or is drawn towards some centre, such as we suppose a planet is towards the Sun, were carried in an ellipse (or some other conic section) by a harmonic circulation, and the centre of the attraction as well as of the circulation is at the focus of the ellipse, the attractions ôr solicitations of gravity will be directly as the squares of the circulations, ôr inversely as the squares of the radii ôr distances from the focus. We find this as follows by a not inelegant specimen of our differential calculus, or analysis of infinites. Let $A\Omega$ be q; BF[15] be e; BE be b (that is, $\sqrt{[qq - ee]}$); the radius $\odot_2M$ be r; $\odot\varphi$ (ôr $\odot_2M - F_2M$) will be $2r - q$, ôr p for short; and let the latus rectum WX be a, which equals $bb{:}q$. Supposing the latus rectum to be a, and representing the constant element of time by θ, let twice the element of area ôr twice the triangle ${}_1M_2M\odot$, which is constant, be θa; and the circulation ${}_2D_3M$ will be $\theta a{:}r$ (see 12 above). Next let us call the difference ${}_2D_2M$ between the radii dr, and the difference of the differences ddr. Now, by the preceding, dr (ôr ${}_2D_2M$) is to $\theta a{:}r$ (ôr ${}_2D_3M$) as $\sqrt{[ee - pp]}$ to b. Therefore $brdr = \theta a\sqrt{[ee - pp]}$, which is the differential equation. But the

[13] This should read 'the straight line ${}_3MR$', as Leibniz corrects it in 1706; see chapter 24.

[14] This should be '$\odot H$, H_3M and ${}_3M\odot$', as Leibniz corrects it in 1706; see chapter 24.

[15] This is a misprint, as Leibniz notes in 1706: BF should be $\odot F$, the distance e between the foci; see chapter 24.

differo-differential equation of this (according to the laws of calculus explained by us elsewhere in these *Acta*) is $bdrdr + brddr = 2pa\theta dr{:}\sqrt{[ee - pp]}$. From these two equations one needs to eliminate dr so that only ddr remains, giving $ddr = bbaa\theta\theta - 2aaqr\theta\theta{:}bbr^3$,[16] from which what was proposed follows. For the element of paracentric velocity ddr is the difference between $bbaa\theta\theta{:}bbr^3$, that is, $aa\theta\theta{:}r^3$, which is twice the centrifugal endeavour (by 12 above), and $2aaqr\theta\theta{:}bbr^3$, that is, since $bb{:}q = a$, $2a\theta\theta{:}rr$. Therefore (by 15) $2a\theta\theta{:}rr$ must be the solicitation of gravity, which, multiplied by [91/92] the constant $a/2$ gives $aa\theta\theta{:}rr$, the square of the circulation. Therefore, the solicitations of gravity are directly as the squares of the circulation, and hence inversely as the squares of the radii. The same conclusion follows both in the hyperbola and the parabola, but above all in the circle, which is the simplest ellipse. The reason for the difference among these conic sections, and for why the circles and ellipses are generated in preference to the other conics, will appear below.

(20) *The same planet is attracted by the Sun in different ways, and in fact in proportion to the square of its proximity*; so that the same one is perpetually solicited to descend towards the Sun by a new impression that is four times as strong when it is twice as near, and nine times as strong when it is three times as near. This is clear from the preceding, supposing the planet describes an ellipse, and circulates harmonically, and moreover is continually impelled towards the Sun. I see this proposition has already been made known also by the most celebrated Isaac Newton, as appears from a review in the *Acta*, although I am not able to judge from that how he arrived at it.[17]

(21) It is also clear that *the solicitation of gravity on the planet is to the centrifugal endeavour of the planet* (ôr the outward endeavour resulting from the harmonic circulation itself grabbing it in its orb and thereby endeavouring to drive it outwards) *as its present distance from the Sun to a quarter of the latus rectum of the planetary ellipse*, namely as r to $a{:}4$, and therefore the ratios of gravity to centrifugal endeavour are proportional to the distances of the planet from the Sun.

[16] In 1706 Leibniz says there should be a comma before the colon between $2aaqr\theta\theta$ and bbr^3, so the formula should read $bbaa\theta\theta - 2aaqr\theta\theta,{:}bbr^3$, or in modern notation, $(b^2a^2\theta^2 - 2a^2qr\theta^2)/b^2r^3$. Since $a = b^2/q$, this gives $ddr = a^2\theta^2/r^3 - (2/a)a^2\theta^2/r^2$, and since $a\theta = dt$, this gives $ddr = (1/r^3 - 2/ar^2)dt^2$.

[17] As noted in the Introduction, Leibniz is not telling the truth here: he did have access to Newton's *Principia* for a while when he was in Vienna in the autumn of 1688, on which he made several notes. Leibniz's own attempts to solve the problem by his own methods in successive manuscripts were made shortly afterwards, as has been demonstrated in detail by Domenico Bertoloni Meli in his (1993). Meli transcribes and annotates all these notes and manuscripts there together with detailed analysis and commentary.

(22) The velocity of a planet around the Sun is everywhere greater than the paracentric velocity, that is the velocity of approaching or receding from the Sun. For since the circulation to the paracentric velocity is as b to $\sqrt{[ee - pp]}$ (by 18, adding the calculation in 19) the former will be greater than the latter if $bb + pp$ is greater than ee, which is certainly the case, since bb is greater than ee, that is to say, b, the transverse axis, is greater than e, the distance between the foci. This in fact always happens in the planetary ellipses known to us, which do not much differ from a circular one.

(23) At *aphelion* A and *perihelion* Ω there is only a *circulation* without approach or recession, *greatest* at perihelion, *least* at aphelion. Moreover, *at the mean distance* of the planet from the Sun (which is at the endpoints of the transverse axis, B and E), the velocity of approach or recession is to the circulation in proportion as the distance between the foci is to the transverse axis, that is, as e to b. For there, p vanishes. [92/93]

(24) *The velocity of the planet's approaching the Sun, or of receding from it,* is *greatest* when the distance $W\odot$ or $X\odot$ of the planet from the Sun is equal to half the latus rectum of the ellipse, for then (by 19 or 21) $ddr = 0$, since $r = a/2$. Thus if from the Sun as a centre, a circle is described with half the latus rectum $\odot W$ as radius, it will intersect the ellipse of the planet at the two points W and X having maximum paracentric velocity, which will be a velocity of approaching at one point, W, and of receding at the other point X. It will be a *minimum* or null at the aphelion and perihelion, that is, at both the vertices of the ellipse A and Ω.

(25) Always in an ellipse, and thus *always in a planet, the centrifugal endeavour* of receding from the Sun, that is, the endeavour outwards *of the harmonic circulation is smaller than the* solicitation of gravity ôr central *attraction of the Sun*. For (by 21) the attraction is to the centrifugal endeavour as the distance from the Sun ôr focus is to a quarter of the latus rectum, but in an ellipse the distance from a focus is always greater than a quarter of the latus rectum.

(26) *The impetuses that a planet receives by the continued attraction of the Sun during its journey are as the angles of circulation*, that is, the angles comprised by the radii drawn from the Sun to the initial and final points of its path, ôr as the apparent motion ôr path seen from the Sun. Thus the impetus impressed during the path A_1M is to the impetus impressed during the path A_1M as the angle $A_1\odot M$ to the angle $A_3\odot M$. For the increments of the angles are as the impressions of gravity (by 17 and 19), so their respective sums will also be proportional, namely the complete angles of circulation to the sums of the impressions ôr impetuses acquired at that point. Hence at the

point *W* where an ordinate from the Sun[18] meets the ellipse, the impetus acquired in going from the aphelion *A* is half the impetus acquired in going from the aphelion to the perihelion; moreover there the distance from the Sun, ⊙*W*, is half the latus rectum. And the impetus acquired in any path is to that acquired in half a revolution as the angle of circulation to two right angles. However, I understand the impetuses impressed by gravity ôr attraction solely by themselves, not with the contrary impetuses impressed by the outwards endeavour subtracted or computed from them.

(27) But it is worthwhile *to explain* more distinctly from the assigned causes *the whole revolution of the planet, and the degree of approach or recession with respect to the Sun*. Thus a planet placed at the greatest digression *A*, ôr at aphelion, experiences both a centrifugal endeavour of circulation driving it outward [93/94] and an attractive solicitation of gravity that are even smaller than if it were nearer the Sun. At that distance, however, namely at the vertex farther from the Sun, the gravity is stronger than twice the centrifugal endeavour (by 21), because the distance ⊙*A* of the aphelion ôr more distant vertex from the Sun ôr focus is greater than half the latus rectum ⊙*W*. Thus the planet descends towards the Sun by the path *AMEW*Ω, and the impetus continually increases while it is descending, as in accelerating heavy bodies, while the new solicitation of gravity remains stronger than twice the new centrifugal endeavour; as long as this happens, the impression of approaching increases over the impression of receding, and so the velocity of approach increases absolutely, until it comes to the place where those two new contrary impressions are equal, that is, at the place *W*, where the distance from the Sun ⊙*W* equals half the latus rectum. At that point, therefore, the velocity of approach is a maximum, and it stops increasing (by 24). From then on, however, even though the planet continues to approach the Sun, the velocity of approach decreases again until the point Ω, with double the centrifugal endeavour prevailing over the impression of gravity. And this continues until the centrifugal impressions from the beginning *A* up to this point, collected into one, precisely consume the impressions of gravity from the beginning to this point, also collected into one, that is to say, when the whole impetus of receding (acquired from the individual centrifugal impressions collected together) finally equals the whole impetus of approaching (acquired from the continually repeated impressions of gravity), where all approaching ceases; and this place is the perihelion itself, at which the planet is

[18] Here it should be remembered that in Leibniz's time it was customary to draw ordinates horizontally and abscissas vertically.

maximally close to the Sun. Afterwards, however, with its motion continuing, whereas up to this point it will have been approaching, it now begins to recede, and tends from Ω through *X* towards *A*. For twice the centrifugal endeavour, which had begun to prevail over the impression of gravity from *W* to Ω, still continues to prevail from Ω to *X*. And therefore since the planet begins to move anew from Ω, with the previous contrary impetuses cancelling each other out, a recession also prevails from Ω, and the velocity of recession continually increases up to *X*; but its increment ôr new impression decreases until that new impression of recession, ôr twice the centrifugal endeavour, is equal to the new impression of [approaching][19] ôr of gravitation, namely at *X*. Thus the maximum velocity of recession is at *X*. And from this point on, gravity ôr the new impression of approaching prevails, although the whole impetus of receding, that is the sum of all the impression of receding acquired since Ω, prevails for quite a long time [94/95] over the whole impetus of approaching impressed anew from Ω. But since after *X* the latter increases more than the former, they at last become equal at *A*, where they cancel each other out, and the recession ceases, that is, it returns to the aphelion *A*. And so with all the original impressions consumed by the compensating equal contrary ones, the thing returns to its initial state; and the entire thing will be repeated in perpetual plays, until one far off day, with time's circuit complete, it brings about a notable change in the constitution of things.

(28) We have then *in the elliptical motion of a planet six especially noteworthy points*. Four, indeed, are obvious: the aphelion *A* and the perihelion Ω, and likewise *E* and *B* at the mean distance (for ⊙*E* or ⊙*B* is half the major axis *A*Ω, and thus the arithmetic mean between the maximum digression ⊙*A*, and the minimum, ⊙Ω). We have added two others, *W* and *X*, the endpoints of the latus rectum *WX*, the straight line applied as an ordinate of the axis at the focus ⊙, which are the points of maximum velocity, at *W* the velocity of recession, at *X* that of approaching (by 24). Here too (by 26) the impetus acquired by the continual impression of gravity from *A to* W is precisely half what is acquired by the whole descent from *A* to Ω; similarly, that acquired by the descent from Ω to *X* is half what is acquired from Ω to *A*; and in all, the impetuses acquired by gravity through *AW*, *W*Ω, Ω*X*, and *XA* are equal.

(29) Now it is time to treat *the causes which determine the type of planetary ellipse*. The focus of the ellipse is ⊙, which is the place of the Sun. Now given the place *A* where the planet begins to be attracted by the Sun, for instance at the planet's maximum distance from the Sun, the vertex that is farther from

[19] The text has 'of recession', which is clearly a slip of the pen.

this focus is also given. Moreover, given the ratio of gravity—that is to say, the virtue by which the Sun begins to attract the planet—to the centrifugal endeavour by which the circulation there strives to drive the planet outwards and away from the Sun, this also gives the principal latus rectum of the ellipse *WX*, that is, the ordinate at the focus ☉. For the given ☉*A* is to the semi-latus rectum ☉*W* in the given ratio of the solar attraction to twice the centrifugal endeavour. But if now a quarter of the latus rectum is subtracted from the maximum given digression ☉*A*, the remainder will be to ☉*A* as ☉*A* to *AΩ*: therefore, the major axis of the ellipse *AΩ* is given, as is its transverse axis. Therefore, given the points ☉, *A*, *W*, and *X*, the point Ω is also given, and hence moreover also *C*, the centre of the ellipse, and the other focus *F*, and the transverse axis *BE*, and thus the ellipse itself. And all these things would be no less given if the attraction began at Ω instead of *A*. [95/96]

(30) From these considerations it is at the same time clear *how an ellipse or* (what is included in that category) *a circle, and not another conic section, is described by the planets.* And a circle, indeed, results when the attraction of gravity and twice the centrifugal endeavour arising from the beginning of the attraction are equal; for then they will remain equal, since there exists no cause for approaching or receding. But when the attraction and twice the centrifugal endeavour are unequal at the beginning (or in a state where the previous opposed impetuses of approaching or receding cancel each other, which is equivalent to the beginning, that is, at aphelion or perihelion), provided that (by 25) the single centrifugal endeavour is smaller than the attraction, an *ellipse* is described; and with the attraction prevailing, the beginning is the aphelion, whereas if twice the centrifugal endeavour prevails, it is the perihelion. If the single centrifugal endeavour is equal to the attraction there results a parabola if it is greater, a hyperbola whose inner focus is the Sun. But if a planet were endowed with levity and not gravity, and were not attracted by the Sun but repelled by it, an *opposite hyperbola* would arise, namely, one whose outer focus would be the Sun itself.

Two especially important points now remain to be accounted for in this argument: one, to explain how the motion of the aether makes the planets heavy, that is to say, drives them towards the Sun, and indeed in the ratio of the squares of their proximity to it; second, to explain what is the cause of the relationship between the motion of the different planets in the same system such that the periodic times are in the sesquialterate ratio of the mean distances, ôr what comes down to the same thing, of the major axes of the ellipses; that is, the motion of the solar vortex ôr aether constituting each system should be explained more distinctly. But since these matters must be re-examined more deeply, they cannot be included in the short compass of this essay; and what seems fitting to us will be expounded more appropriately on a separate occasion.

11

Observations on the Cause of Gravity and its Properties[1]

Denis Papin, *Acta eruditorum*, April 1689[2]

In order to provide further confirmation of the usefulness of very large tubes and balances, which I have already tried to provide by means of several examples, I have decided to propose the Machine invented by Mr Perrault, and published by Mr Blondel in the last chapter of the book he wrote in France,[3] concerning the projection of iron balls filled with gunpowder. For on calculation it can evidently be demonstrated that the said machine, other things being equal, will happen to produce a much greater effect if one employs equal tubes for impressing motion on them, rather than common weights; a truth, according to the evidence of our calculation—I do not judge it to be so from the facts—that would shine forth more easily if I were to put forward a few things concerning the nature and properties of gravity, up until now scarcely known.

Galileo, in his *Dialogues on Mechanics*, supposes that *to a falling heavy body there accrue* [183/184] *equal degrees of velocity in equal times*; and Mr Blondel, in the above cited work on pp. 453–477, tried to evince the same truth a

[1] From the Latin: '*D. Papini De Gravitatis Causa et proprietatibus Observationes*', *Acta eruditorum*, April 1689, 183–8. Translated by Richard T. W. Arthur.

[2] Leibniz had known Denis Papin since his time in Paris, when (between 1673 and 1675) Papin had served as a laboratory assistant to Christiaan Huygens at the *Académie des Sciences*. As someone committed to a Cartesian framework in natural philosophy, it was natural for Papin to take it on himself to reply to Leibniz's attack in the *Brevis demonstratio*. He did not find occasion to do so, however, until three years after its publication, when he was settled with a professorship in mathematics at Marburg. He defends Descartes's theory of gravity against criticisms made in the 1686 *Acta* by Johann Christoph Sturm and Jakob Bernoulli, as well as by Leibniz, appealing to what he takes to be implicit in as yet unpublished experiments by Huygens. See O'Hara (forthcoming, ch. 1) for discussion.

[3] Here Papin refers to François Blondel's *L'Art de jetter les bombes*. Paris, 1683. Perrault's machine is a device for projecting grenades or little bombs into the enemy's positions. It is described (with an illustration) by Blondel in Part III, chapter 3, of his book, pp. 419 ff.: 'It is a wheel or drum *A* around which is coiled a cord carrying the weight *B*, the bar *CD* passes behind the drum and is attached to its pivot *E*, so that the weight *B*, turning the drum by its fall, also gives motion to the bar, which describing the arc of the circle *CI* and striking against another solid pivot and well attached to the point *F*, makes the ball *G* go with the speed the motion of the ball impresses on it, along the straight line *FH* which touches the arc *CF* at the point *F*.'

Leibniz: Journal Articles on Natural Philosophy. Richard T. W. Arthur, Oxford University Press.
DOI: 10.1093/oso/9780192843531.003.0012

posteriori by many experiments, as if no a priori demonstration could be given. In fact, the most illustrious Huygens confirmed the matter using an outstanding method: for this proposition, which others assume as a principle, he demonstrated a priori by means of another truth, which goes as follows. *The power that is the cause of gravity has a speed that is infinite in comparison with the velocities of the falling bodies that may be observed by us.* For with this principle supposed, Galileo's proposition quoted above follows of its own accord: and, indeed, if each speed acquired by a falling body is to be judged infinitely slow in comparison with the speed of the moving power, then it is clear enough that the said heavy body, whether it is at rest or moving, ought to be affected by the moving power always in the same way; since an infinitely slow motion cannot be distinguished from rest by any sensible effect. And therefore there is no reason why the cause of gravity would not impress the same quantity of motion in a given time as it had first impressed; in succeeding equal moments it would continue in a curve in the same tenor. It therefore remains for it to be demonstrated through experience that the velocity which we have ascribed to the efficient cause of gravity is infinite. Indeed, experiments for evincing this should be set up in various ways in accordance with the variety of opinions which are proposed for explaining the cause of gravity.

But in fact, putting aside the opinions of others who misuse their works in order to explain the obscure in terms of the equally obscure, the most distinguished Huygens, following only the Cartesian hypothesis, has undertaken to confirm this infinite speed by experiment.[4] To this end he prepared a board constructed in such a way that it could be rotated horizontally around its own centre.[5] The centre was pierced in such a way that it would allow a thread to be passed through it, with two equal globes attached to its ends, one of which was kept hanging under the board along the axis. The other globe was lying on the board in such a way that it was immediately drawn towards the centre by the weight of the first globe unless some force restrained it. Then the board was rotated very fast about its axis. Now when the globe lying on the board attained a certain velocity it was observed that it tended to recede from the centre of its motion with so much force that it could resist the weight of the other globe hanging under the table; [184/185] for then the hanging globe did not descend, even though it was held up by no other power but

[4] Here Papin refers to the explanation of gravity that Christiaan Huygens is about to publish in his *Discours de la Cause de la Pesanteur*, 1690.

[5] 'J'ay fait avoir cy devant cette mesme proprieté de mouvement circulaire, en attachant des corps pesants sur une table ronde, percée au centre, & qui tournoit sur un pivot, & j'ay trouvé la determination de la force...' Huygens, *Discours de la cause de la pesanteur* (1690, 130).

the centrifugal force of the globe rotating with the board. So, by then measuring the circumference described by that globe, he was able to show through experience how great a velocity in a given circumference would have a centrifugal force equal to the force of gravity. Furthermore, Huygens demonstrated that in large circumferences the speed required is greater than that required in small ones in order to have an equal centrifugal force in each. In the circumferences of greater circles on the surface of the Earth, which exceed the circumferences described on the board by such an enormous distance, it is therefore necessary that the matter causing gravity be moved extremely fast so that it could have so much centrifugal force. For that matter only pushes heavy bodies down towards the centre of the Earth with the same force with which they endeavour to recede from the same centre; as in fact, by having duly performed the calculation, the same most distinguished Huygens discovered that the matter causing gravity would have to move with such a great impetus that it would be able to traverse the whole perimeter of the Earth almost a thousand times in any single hour. But such a velocity can very justifiably be regarded as infinite with respect to the speeds acquired by falling bodies.

Having thus demonstrated the extremely rapid motion of the matter which is the cause of gravity, and having traced the issue back to the first principles themselves, the greatest light will be shed on resolving difficulties which seemed at first sight to be insuperable. For example, I have seen many people asserting that according to the Cartesian view heavy bodies ought to descend towards the Earth's axis, and only bodies situated at the equator would tend towards the centre—(you can see how this problem was aired by Sturm and Bernoulli already some time ago in the *Acta eruditorum* of February 1686)[6]—and, of course, if the diurnal motion of the Earth about its axis would suffice for causing gravity, then the validity of the objection could not at all be denied, and all heavy bodies would tend towards each of the parallels' own centre. But, in fact, that motion is so slow in comparison with the velocity of the matter causing gravity that the parts on the surface of the Earth that are moving with the greatest velocity, namely those on the equator, could themselves be regarded as being unmoved in relation to the extremely rapid motion of the matter causing gravity. For the equator revolves only once in 24 hours, whereas that matter can complete 24,000 circuits [185/186] in the same time. No wonder, then, if that diurnal motion produces no sensible effect by

[6] '*Dubium circa Causam Gravitatis a rotatione Vorticis terreni petitam*' ['A Doubt concerning the Cause of Gravity derived from the rotation of a terrestrial Vortex'], *Acta eruditorum*, (Bernoulli 1686). In the article Jacob Bernoulli describes his exchanges with Johann Christoph Sturm on this issue.

altering the descent of heavy bodies, and instead they all descend towards the centre, no differently than if the Earth were at rest.

And it is just as easy to resolve another difficulty selected from the properties of ascending heavy bodies, the one proposed by the most distinguished Leibniz in the *Acta eruditorum* of May 1686, where he strives to demonstrate against Descartes that the motive force in bodies is different from the quantity of motion. For when double the quantity of motion is given, it is found that the motive force is quadrupled in the same thing; inasmuch as with double the quantity of motion a heavy body could ascend to four times the height. And, of course, it is established by Galileo's experiments that any heavy body, when it is moved upwards with two degrees of velocity, can ascend to a height of four times that to which it would ascend if there were only one degree of velocity. The validity of this objection would also be indisputable if the spaces traversed in ascending had to be, as our author reckons, proportional to the motive forces. But, in fact, from what was said above the response is easy: for it is known that in these discussions we have abstracted away from all resistances and impediments, which do not arise from lifting the body under gravity; since, however, the velocity of the matter causing gravity is infinite, it follows that it is always equally efficacious, by whatever motion the heavy bodies are carried. And, therefore, if they ascend faster, they will receive neither more blows nor more powerful ones from the said matter than if they moved upwards much more slowly in the same time. Therefore, in equal times the motive force of those things will no more press them down while they go upwards through large spaces than while they go through small ones. So the most distinguished Leibniz is wrong to estimate that the motive forces in this case should be proportional to the spaces traversed by the ascending body; since, on the contrary, the times of ascent and the motive forces ought to be in the same ratio, so that the individual parts of the motive forces should be subtracted from the individual parts of the times: but the times are in the ratio of the square-roots of the spaces, so the motive forces will be too.

But if he rejoins, however, that in having acknowledged it to be the case that the motive force which raises 4 pounds to a height of 1 foot is the same as the force which raises 1 pound to a height of 4 feet, it follows that the spaces traversed in ascending and the motive forces will be in the same ratio; and no mention of time [186/187] customarily occurs in determining motive force; I respond as follows: That holds in machines where a new force is always impressed on the body. For there the time is equally increased or diminished in proportion to both the motive force and to the resistance. Here, however, the ratio is very different, for the ascending body already has in itself sufficient force, and does not acquire any new one; and so the only ratio to be had is to the resistance, which always takes away some force. But resistance does not depend on the quantity of space traversed, as we have shown above, but on the quantity of time. These things are too easy for us to dwell on them at great length. But although here I seem to be defending Descartes's position, I would not want the most distinguished Leibniz to think that for this reason I likewise wish to defend the remaining views of Descartes concerning the laws and rules of motion. If, then, it seems that something in what I have written here should be disapproved, I will try to remove any scruples; but if he objects to other things, I do not believe any of that pertains to me.

Meanwhile I shall proceed to remove further scruples that could remain about the Cartesian hypothesis. The most serious doubt about the Cartesian system was proposed by the celebrated author of [the article in] the *Nouvelles de la République des Lettres* of December, 1685, p. 1317.[7] 'It does not seem,' he says, 'that the matter of the first element, from which the Sun is composed, could impress a new force on the globules of the second element towards the circumference; for if it could, then it would have a greater force for receding from the centre than the said globules. And so it would in fact recede from it, and would impel the remaining bodies towards the said centre, but not towards the circumference of the vortex.' Thus, it seems the Cartesian system could in no way remain in being, inasmuch as the matter of the Sun should immediately be scattered and dispersed in every direction. By a similar argument it could be proved that the terrestrial vortex should for a short time become mixed up with the Sun's vortex. For since the matter contained in the smaller vortex would describe smaller circumferences, it follows that, as was said above, its centrifugal force ought to be greater than the centrifugal force of the neighbouring matter constituting the larger vortex. And so nothing can prevent the matter of the smaller vortex from continually scattering and immediately becoming mixed up with the larger vortex.

In order to solve this difficulty, I believe we should have recourse to the congruity and incongruity of parts, which produce many important effects in observable fluids. For what would prevent us from believing that there is also in other very subtle [187/188] fluids a congruity or incongruity of parts, and that the effects of those qualities would be found to be much greater in more subtle liquids than in dense ones? Their force is certainly situated in the

[7] We are unable to verify this reference. On p. 1317 of the *Nouvelles de la République des Lettres* of December, 1685 there appears an article '*Dialogues entre Photin et Irenée sur le dessein de la réunion des Religions & sur la question si l'on doit emploier les peines & les recompenses pour convertir les Hérétiques*', which is clearly not the one Papin is quoting from.

surface of bodies; but surfaces, considered in relation to volume, are greater in smaller bodies than in larger ones.[8] Therefore just as a tiny drop of oil in the middle of water, on account of the homogeneity of its parts, remains united, and does not get mixed up with the particles of water, being incongruous with them, so it is very probable that the parts of the terrestrial vortex, because of their congruity, remain united with one another, and cannot become mixed up with the matter of the large vortex, being incongruous with them. In this way I hold that the fixity of liquefied gold proceeds only from the congruity of parts; for so much cohesion can arise from this that it cannot be overcome by the greatest impulse of the particles of fire. Thus, it is also just as probable that the matter of the Sun, by however much force it strives to scatter in every direction, still cannot be torn to pieces because of the adhesion arising from the congruity of its particles. With this, then, I regard myself as having satisfied the objection of Achilles.

Perhaps, however, someone would also ask me to explain more clearly what this congruity of parts is that so much adhesion in a curve should arise from it? In a word, I would say that the thing is extremely evident in observable bodies. For we see hairy bodies with spikes adhere together more than with polished bodies; and polished bodies, on the other hand, adhere better with other polished ones than with hairy ones. For there is greater contiguity of one polished body with another than of a polished one with a hairy one. But contiguity necessarily implies adhesion. For as long as bodies are contiguous they have their external surfaces exposed to the pressure of the ambient fluids, by which they are impelled together and united; but their internal surfaces, since they are contiguous, do not admit any matter between them which could drive them apart from one another.

Anyone who desires to see a great many experiments established about the congruity and incongruity of parts, should consult the *Philosophical Transactions of London* of 1674 or 1675. For I remember that many things were communicated about this argument at that time by the most illustrious Mr Boyle, and he indicated at the same time that such things would perhaps be of some use in explaining the great system of the world.[9]

[8] That is, since the volume of a sphere increases as r^3 and its surface area as r^2, a force distributed across its surface will be stronger in inverse proportion to its radius.

[9] Papin is probably referring to 'Of the Excellency and Grounds of the Corpuscular or Mechanical Philosophy' (Boyle 1674).

12

On the Isochronous Line along which a heavy body descends without acceleration, and on the controversy with the Abbé D. C.[1]

G. W. Leibniz, *Acta eruditorum*, April 1689[2]

In the March 1686 issue of these *Acta* I published a demonstration against the Cartesians, in which a true estimation of forces is presented, and it is shown that what is conserved is not quantity of motion, but quantity of power, which is different from quantity of motion. To this a certain learned man in France, the Abbé D. C., responded on behalf of the Cartesians, but, as became clear later, without having sufficiently perceived the force of my argument. For he believed that I was attacking certain other accepted principles, which he enumerates in the *Nouvelles de la République des Lettres* of June 1687, p. 579, and on pp. 579 ff.[3] refuses to acknowledge the contradiction which it seems to me I had found in them; when in fact it had never [195/196] entered my mind to doubt them, as I advised in the *Nouvelles de la République des Lettres* of September 1687. Likewise, in order to evade my objection, he

[1] From the Latin: *G. G. L. De Linea Isochrona in qua Grave sine Acceleratione Descendit et de Controversia cum Dn. Abate D. C., Acta eruditorum*, April 1689, 195–8. (ESP 173–77; GM V 234–37—Gerhardt erroneously expands '*Abate D. C.*' to '*Abate De Conti*' instead of 'De Catelan'.) Translated by Richard T. W. Arthur.

[2] In this article Leibniz provides the solution to the challenge problem he had proposed to Catelan in his second reply, that of finding the isochronous line—the curve traced by a heavy body falling under gravity in such a way that its velocity vertically downwards remains constant. Although he had worked out his solution when he proposed the problem, he did not publish it until he was on his travels in Italy. Using his calculus he has calculated the equation of the curve as $y^2aP = z^3$, but in this paper he does not reveal his calculations, instead expressing his results geometrically, serving as a lesson to the Cartesian mathematicians. Since Huygens had already solved this problem, Leibniz offers them a further, related challenge problem, that of finding the curve along which a descending heavy body recedes from or approaches a given point uniformly.

[3] Here there was a misprint, 'p. 519' instead of 'p. 579'.

Leibniz: Journal Articles on Natural Philosophy. Richard T. W. Arthur, Oxford University Press.
DOI: 10.1093/oso/9780192843531.003.0013

launched himself into a digression which, according to the way I conceived the state of the controversy, was accidental to it. For, if the heights remain the same, the same force is acquired or lost by falling bodies whatever time is allotted, and this time is increased or diminished in proportion to whether the inclination is greater or less.

Using this opportunity, and in order to make clearer that the time, and thus the distinction between isochronous and non-isochronous powers, has nothing to do with the matter, and also so as to gain some increase in knowledge from our dispute, I posed the following problem for him in the said *Nouvelles* of September 1687, whose solution, seemingly quite an elegant one, is to be written up by me in the meantime: *To find the isochronous line along which a heavy body uniformly descends, that is to say, approaches the horizontal by equal amounts in equal times, and is thus carried downwards without acceleration and with a velocity that always remains constant.*

But the Abbé D. C. has not sent anything further in reply, whether because he did not want to handle the problem, or because, having finally come round to my point of view, he judged himself satisfied. But in his stead the most celebrated Christiaan Huygens judged the problem worthy of his attention, and his solution, very consonant with my own, is given in the *Nouvelles* of October 1687, although he has suppressed the demonstration and explanation of the difference between the lines of the same kind, as he calls them, which, he notes, satisfy the problem.[4] So I want to provide these here, which I would have done all the sooner, if I had not been expecting some fruit from the labours of the Abbé.

Problem: To find the plane line along which a heavy body descends without accelerating. [Fig. 12.1]

Solution: Let there be a quadrato-cubic paraboloid βNe (that is to say, one where the product of the square of the base NM by the parameter aP is equal to the cube of the height βM) situated in such a way that the tangent βM at the vertex β is perpendicular to the horizontal.[5] If at any point N of the line, the

[4] Christiaan Huygens, '*Solution du problème proposé par M. L[eibniz]*' (Huygens 1687).

[5] Thus the equation of the curve is $y^2aP = z^3$, where the ordinate y represents the base NM, and z is the height βM and aP the parameter. Here $z = 4x - a$, where x is the original abscissa and a is a constant. Characteristically, Leibniz just provides the result, without the calculation. A manuscript containing the calculation, however, was attached to his copy, and was published by Gerhardt (GM V 241). Leibniz assumes that in free fall the time of fall is proportional to the square root of the vertical height x through which the body has fallen, so that $t^2 = ax$, where a is a constant of proportionality, giving an element of time equal to $\frac{1}{2}(ax)^{-1/2}\, a\, dx$, or $adx/2\sqrt{(ax)}$. Now 'since, with the speeds equal, the times will be as the spaces', the time dt in which the body traces an element of the curve ds is to the time it traces a vertical element dx in free fall as $\sqrt{(dx^2 + dy^2)}$ to dx, giving $dt : adx/2\sqrt{(ax)} = \sqrt{(dx^2 + dy^2)}:dx$, or $2dt\, dx\sqrt{(ax)} = adx\,\sqrt{(dx^2 + dy^2)}$. But 'the elements of times of

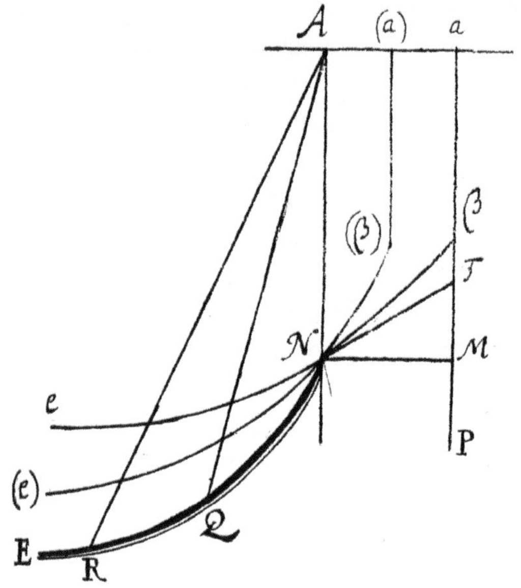

Fig. 12.1

heavy body by descending is supposed to be endowed with the speed which it could acquire by descending from the horizontal *Aa* whose height $a\beta$ above the vertex β is $^4/_9$ of the parameter of the curve, then the same heavy body will then descend uniformly along the line *Ne*, however far it is continued, as was desired.

Demonstration: Let the straight line *NT* be tangent to the curve βNe at *N*, intersecting βM at *T*. In every case (by a known property of the tangents to this [196/197] curve), *TM* to *NM* will be in the subduplicate ratio of $a\beta$ to βM.[6] Therefore *TM* to *TN* will be in the subduplicate ratio of $a\beta$ to $a\beta + \beta M$, that is, to *aM*. Now, the ratio of *TM* to *TN* is equal to the velocity of descent which the heavy body supposed at *N* possesses *along the curve* (that is, the velocity with which it thereafter approaches the horizontal while describing the curve) to the velocity with which the same body would then descend from *N*, not along the curve, but *in free fall*, if it could (as is known from the nature of the inclined plane). But then this velocity of free fall is to a certain constant in the

descent dt will be proportional to the elements of the descent dx' giving $2dx\sqrt{(ax)} = a\sqrt{(dx^2 + dy^2)}$, or $dy^2 = dx^2(4ax - a^2)/a^2$. Changing the variable to $z = 4x - a$, we get $dy^2 = dz^2(z/16a)$. So $dy/dz = (1/4)\sqrt{(z/a)}$, giving $y = (1/36a)\, z^3$, so the parameter $aP = (1/36a)$.

[6] That is, $TM{:}NM = \sqrt{a\beta} : \sqrt{\beta M}$. Here $TM = dx$ and $NM = dy$ in the characteristic triangle *NTM*. The calculation of the 'known property of the tangents to this curve' is as follows: since $y^2 aP = x^3$, we have $2ydy\, aP = 3x^2 dx$, giving $dx/dy = 2y\, aP/3x^2 = (^2/_3)\sqrt{(aP/x)}$. But $a\beta = {}^4/_9\, aP$, so $TM/NM = dx/dy = \sqrt{(a\beta/x)} = \sqrt{(a\beta/\beta M)}$.

subduplicate ratio of aM to $a\beta$, since the velocities in free fall are (as is known from the motion of heavy bodies) in the subduplicate ratio of the heights (from which they are produced by falling).[7] Therefore the velocity of descent along the curve *Ne* possessed by a heavy body at any supposed point *N* of the curve, is to a constant velocity in the compound subduplicate ratio of $a\beta$ to aM and aM to $a\beta$, which is a ratio of equality. Therefore, this velocity of descent along the curve is itself constant, that is, everywhere the same along the curve *Ne*. Which is what was to be shown.

Consequences: (1) A heavy body having fallen from a height of *Aa* can descend from the same point *N* by infinitely many isochronous curves, but all of the same species, that is, differing only in the magnitude of the parameter—for example, *Ne*, *N*(*e*), and *NE*—and these curves are all quadrato-cubic paraboloids, and thus similar to each other. Moreover, each paraboloid of this kind will serve here, provided they are so disposed that the distance $a\beta$ or $(a)(\beta)$ from the horizontal $a(a)$ from which the heavy body began its fall, is $^4/_9$ of the parameter βe or $(\beta)(e)$ of the curve. It does not matter whether the heavy body that will descend along the isochronous curve *N*(*e*) has reached *N* from *a* (*a*) through the path $(a)(\beta)N$ or by another route, or even whether it has acquired the same speed and the same direction due to some other cause without having fallen at all. But of all these isochronous lines along which the heavy body can descend from *N* without acceleration, that which offers the fastest descent with the point *N* as its vertex, is *NE*, whose tangent is the straight line *AN* perpendicular to the horizontal.

(2) The time of descent along the straight line $a\beta$ is to the time of descent along the curve βN as half the height βM to $a\beta$ itself. And so if βM is twice $a\beta$, the times of descent through $a\beta$ and βN will be equal. The reason for this is evident: for the times of a uniform descent are in the ratio of the heights, and, as Galileo has demonstrated, [197/198] the time in which a moving body traverses a height $a\beta$ by an accelerated motion is twice that in which it traverses an equal height βM (as happens here, albeit through the curve βN) by a uniform motion having a speed equal to the final speed acquired at β through an acceleration.

I confess that I did not compose this problem for geometers of the first rank who are proficient in a deeper sort of analysis, but rather for those who share the sentiments of *that erudite Frenchman* who seemed to be offended by my quarrel with the majority of today's Cartesians (who paraphrase their master

[7] That is, the ratio of velocity in free fall vertically downwards to velocity along the inclined plane is as the square root of their heights, i.e. $\sqrt{aM}$ to $\sqrt{a\beta}$.

rather than emulate him). For such people, apart from accepting the dogmas current among the Cartesians, also attribute too much to an analysis in widespread use among themselves, to the point of believing that they can establish anything in mathematics by means of it (if only, that is, they wanted to take the trouble of calculating it)—and this is not without detriment to the sciences, which are less carefully cultivated when there is a misplaced faith in discoveries already made. With this I had wanted to provide material for them to practise their analysis on this problem, which requires not a lengthy calculation, but some finesse.

If any of them should complain that the solution has already been pre-empted, they could try to find *another isochronous line* akin to this one, in which a heavy body uniformly recedes from (or approaches) not the horizontal, as heretofore, but a certain point. Hence the problem would be this: *to find the line along which a descending heavy body equably recedes from a given point, or approaches it.*[8] Such would be a line *NQR* if its nature were such that, from a given or fixed point *A*, with any straight lines drawn to the curve, such as *AN*, *AQ*, *AR*, the excess of *AR* over *AQ* would be to the excess of *AQ* over *AN* as the time of descent along the arc *QR* to the time of descent along the arc *NQ*.

[8] This problem was not solved until June 1694, when Jacob Bernoulli published a solution in the *Acta eruditorum*, '*Solutio Problematis Leibnitiani*...' (Bernoulli 1694b), whose full title translates as 'A Solution of the Leibnizian Problem of the Curve of Equable Approach to and Recession from a given point, mediated by the rectification of the Elastic Curve'. Leibniz gave his own solution the following August in his *Constructio propria Problematis de Curva isochrona paracentrica, ubi et generaliora quaedam de natura et calculo differentiali osculatorum, et de constructione linearum transcendentium*... [Construction of the Problem of the Paracentric Isochronous Curve, together with More General Considerations on the Nature and Differential Calculus of Osculations, and on the construction of Transcendental Curves...] (Leibniz 1694; GM V 309–18).

13

On the Cause of Gravity, and defence against the Cartesians of the author's own view on the true laws of nature[1]

G. W. Leibniz, *Acta eruditorum*, May 1690[2]

When I demonstrated in these *Acta* that the same quantity of motion is not always conserved, and that another law of nature must be established, the Abbé D. C. [De Catelan] endeavoured to respond. But to my mind he has not understood, and has imputed to me opinions which I was far from holding. When this was brought to my attention, I thought I should remain silent, whether because he acknowledged himself satisfied (which, however, it would have been fair to declare publicly), or, not to mention other reasons, because perhaps I did not want to touch on a certain problem [228/229] proposed by me on the occasion of our controversy, which afterwards the most celebrated Huygens solved exactly as it had been posed by me. However, I recently gave in these *Acta* a demonstration and amplification of this construction,[3] and lest the Learned Abbé or someone else might complain that a solution was being snatched away from him, I proposed the problem more or less unchanged, so that a Cartesian analysis could be performed on it: *To find the line in a plane in which a falling heavy body recedes from or approaches a given point equally in*

[1] From the Latin: '*G. G. L. De Causa Gravitatis, et defensio sententiæ suæ de veris naturæ legibus contra Cartesianos*', *Acta eruditorum*, May 1690, 228–39; Figs. 3 and 4, Table VII, facing p. 218. (ESP 178–91; GM VI 193–203.) Translated by Richard T. W. Arthur.

[2] Leibniz had found Papin's rejoinder to his *Brevis Demonstratio* in the course of reading through the issues of the *Acta eruditorum* on his return to Germany after his long journey to Italy. He probably composed this response in Augsburg, and sent it to Otto Mencke in Leipzig for publication in the journal at the end of April in 1690.

[3] This was the first challenge problem initiated by Leibniz, the so-called 'isochrone problem' in the *Nouvelles de la République des Lettres* in September 1687. This was the problem of determining the curve along which a body falling under gravity reaches a given endpoint in the same period of time whatever the starting point on the curve. Huygens published his own solution in the same journal in October 1687. Leibniz published his own mathematical solution in the *Acta eruditorum* of April 1689.

Leibniz: Journal Articles on Natural Philosophy. Richard T. W. Arthur, Oxford University Press.
DOI: 10.1093/oso/9780192843531.003.0014

equal times. Meanwhile another erudite Frenchman (the most distinguished gentleman P. [Papin]), standing in for the Abbé, was moved to defend the opinions of the Cartesians in the *Acta eruditorum* of 1689 (pp. 183ff.). Even though he seems not to have touched on the state of the controversy much, I nonetheless wanted to satisfy his doubts too, especially since it will bear on other things that are not to be condemned, which will provide an opportunity for illustrating the argument. Thus he writes as follows: Galileo *supposes* that a falling weight acquires equal degrees of velocity in equal times. Blondel tried to prove this by experiments with his book on bombs,[4] as if there were no a priori *demonstration* of it. But this Huygens had given, by supposing the motion of the matter which moves the weight to be of infinite speed compared to the velocity of falling weights that can be observed by us, so that the weight, whether at the beginning when it is still at rest, or afterwards when it is already moving, is produced in the same way by the mover in proportion to the quantity of impressed motion or increase in velocity, since it can be held to be always at rest by comparison with the mover. Huygens, however, had estimated the magnitude of the velocity of the mover from the Cartesian hypothesis, by supposing gravity to arise when a certain subtle matter is rotated around the Earth and by receding from the centre pushes other bodies down towards the Earth. And so he investigated by experiment how much centrifugal force in the rotation of a small circle would be equipollent to gravity, and from this inferred that in a large circle, such as is measured by the circumference of the Earth, that matter would move with such a great velocity that it could traverse the whole circumference of the Earth many times in only one hour. And these things indeed are beautiful and worthy of Huygens, but what our author adds on his own account are not equally admissible. That is, he thinks that by means of this he can eliminate some grave difficulties that have troubled scholars. To wit, in these *Acta* those most distinguished gentlemen Sturm and Bernoulli judged it worth their mutual consideration to debate [συζητήσει] how on the Cartesian hypothesis it would happen that heavy bodies are not pushed down toward the axis, but rather to the centre.[5] [229/230] Our

[4] François Blondel, *L'Art de jetter les bombes* (1683).

[5] Leibniz is referring to the article (cited by Papin) by Jacob Bernoulli in the February 1686 *Acta*, pp. 91–5, in which Bernoulli reports his exchanges with Johann Christoph Sturm (Bernoulli 1686). He summarizes as follows: '1. There is no reason why the pushing of the globules should occur in a vertical plane rather than in the plane of a parallel circle, since that tangent along which the globule endeavours to recede is common to infinitely many planes. 2. There is no obvious cause why, if the pushing occurs in a vertical plane, it should occur along the line perpendicular to the tangent *ab* rather than along any other line, in that the globule *a*, surrounded on all sides by infinitely many series of globules, would thereupon repel them in various different directions' (94).

author thinks that this difficulty ceases if only it is considered that the velocity of the matter effecting gravity is incomparably greater than the Earth's velocity of rotation itself, for in this way the difference of velocity which is at the equator or some other parallel will be of no moment in comparison with that very great velocity. My argument made against the Cartesians is satisfied just as easily (if we are to believe him). For, since the velocity of what moves the heavy bodies is infinite, it follows from this that it is exactly as if the heavy body is at rest. Consequently, from the beginning the blows impressed in equal times are always the same, so that the forces are as the times, not as the spaces of ascent or descent, as I had in fact estimated. Finally, since the Author of the [article in the] *Nouvelles de la République des Lettres* was worried that the very matter of the *first element* would recede from the Sun by centrifugal force ahead of the globules of the *second element*, and that the Sun would thus be dispersed, and was likewise worried that the Earth's vortex would be dispersed into the Sun's vortex, which of course describes much larger circles, so that the Earth's vortex would be endowed with a correspondingly smaller centrifugal force (other things being equal); he responded by introducing a certain *congruity* or incongruity *of parts*, which would keep together parts that would otherwise fly away from one another. Thus far, him.

To which I reply as follows:

(1) Galileo did not just *suppose* that the motion in heavy bodies is equally accelerated in equal times, but also tried to confirm it by giving reasons and experiments, nor did he come to this opinion by chance.

(2) Our Objector confuses a *demonstration* of some truth with the giving of a cause according to a certain hypothesis; and he perhaps does not perceive well enough the suggestion of Huygens, whose intention (as far as it can be ascertained) in this reasoning we have cited was not to demonstrate that the nature of the acceleration of heavy bodies is what we have said, but that, having supposed it to be such (from the phenomena, perhaps), to explain a probable way in which it could arise.

(3) But we will give an absolute demonstration of its truth a priori in our Dynamics, without advancing any hypothesis, solely by explaining on the basis of common phenomena, that a heavy body is of the same weight in a higher or lower place, which is indubitable in differences of heights that are at any rate rather small in relation to sense.

(4) The *hypothesis* which he calls *Cartesian* is in fact *Keplerian*, even if it was further developed by Descartes. For before everyone else, Kepler discovered that he could sketch the origin of gravity as long as some fluid

consisting of more solid parts [230/231] made to move in a circle and tending to recede from the centre, would push the less solid parts floating in it down towards the centre. This thought of his—just like many others—Descartes put to his own use, pretending to be the author (in his own unpraiseworthy fashion), just as he also took the explanation of the equality of the angles of incidence and reflection by means of the composition of two motions from Kepler's *Supplemental Notes on Witelo*,[6] and learned the true rule of refraction from Snell. And truly, although Descartes was a very great man, by these artifices he lost a great deal of esteem among intelligent commentators.

(5) Even if it is rightly assumed that the velocity of the matter causing gravity is incomparably greater than that of the heavy bodies around us, it is still not necessary to assume this is in order to explain the equable acceleration of heavy bodies, as the Objector seems to reckon, perhaps having not sufficiently perceived Huygens's intention. In order to make this clear, let us imagine (Fig. 3, Table VII [Fig. 13.1]) a very long horizontal tube *TV*, closed at both ends, full of mercury,[7] in which near the end *V* there is at *G* a globe *G* made of glass or some other material, which is less dense or less solid than the mercury, and not broken up by it. If now this cylindrical tube is rotated in the same horizontal plane about the other fixed endpoint *T*, then the mercury trying to recede from the centre, and tending towards *V*, pushes the globe from there and makes it tend towards *T*, without any ascent. Indeed, even if the tube is somewhat inclined to the horizontal, so that the endpoint *T* is lower than *V*, nevertheless with a sufficiently rapid force of circulation it will bring it about that the globe floating in whatever way in the mercury will descend from *V* towards *T*, with a most appropriate representation of gravity. Moreover, the

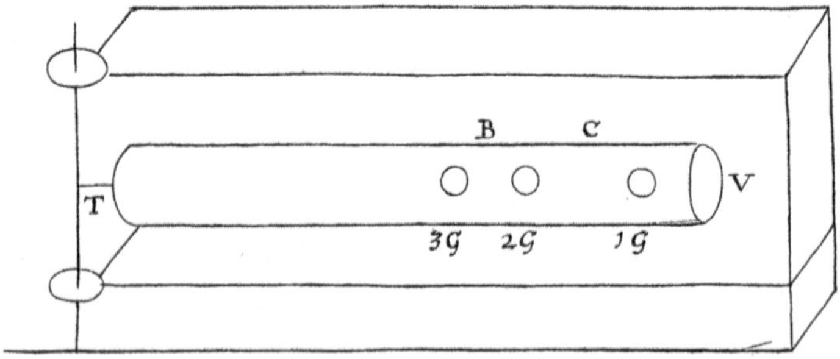

Fig. 13.1

[6] This work by Kepler, *Paralipomena ad Vitellionem*, supplementing and correcting the work on optics of Witelo, Alhazen, and others, was published by him in 1604.

[7] Leibniz initially calls mercury by its formal name, Hydrargyrum.

reason why this centrifugal force of matter receding from the centre pushes other things that are receding less towards the centre, can be distinctly explained in the following way: that the matter *B* (the mercury), receding from the centre *T*, endeavours to insert itself between *C* (the mercury) and the body $_2G$ (the globe), and since the mercury *C* cannot be impelled any further, being obstructed by the cap of the tube, the body $_2G$ is repelled towards *T* ôr to $_2G$. Now with these things supposed, as the speed by which the globe *G* tends towards the centre *T* continually increases while the speed of circulation of the tube decreases, it will happen that somewhere, as at $_2G$, the velocity of the rotating tube will equal that of the globe tending in a straight line towards the centre. Nevertheless, however, if the tube is imagined to be so long that the point $_2G$ is much farther than this from the centre, for example, by many miles, then with the globe tending next from $_2G$ to $_3G$ by an interval [231/232] of many yards, the speed of circulation will not noticeably change, nor will there be a noticeable difference between $_2G$ and $_3G$, and so the centrifugal force by which the globe between $_2G$ and $_3G$ is continually impressed will not noticeably change either. And so it will be exactly as if the globe at *G* remains in the same place and receives equal impressions in equal times. Now, the same thing we have imagined as happening in the tube really happens in the aether, if indeed gravity arises from its centrifugal force. For because of the very great distance from the centre (namely, of the Earth), the extremely small interval during which a heavy body in our experience approaches the centre while falling can make no noticeable difference, and it is therefore from this that an equal impression of speeds arises in equal times, even if the rapidity of the aether is not very great. But if we seek gravity not from the centrifugal force of the circulation, but by the driving of heavy bodies towards the Earth by a certain matter, like a wind, then in that case, in order to explain the increments of gravity being proportional to the times, it is necessary that the speed of the wind be incomparably greater than that which the heavy bodies of our experience acquire.

(6) Even though a while ago I myself favoured deriving gravity from the centrifugal force of the circulating aetherial matter, there are however some things which induce the gravest doubts. And not to mention the other difficulties now, it is necessary that this aether move around the Earth not along the equator or in parallels, but in great circles like the meridians (otherwise heavy bodies will not tend toward the centre but toward the Earth's axis); but in that case it is necessary that the aether be much more concentrated towards the poles; hence it is not clear how heavy bodies are impelled in the same way towards the centre in places on the equator and

those near the poles, which nevertheless happens, and the phenomena teach that no such noticeable difference is observed. If this difficulty could be remedied, it would be easier to believe the cause of gravity thought of by Kepler.

(7) Another cause of the same thing could be assigned that is not susceptible to this difficulty, by conceiving the disposition[8] of a certain matter propelled from the globe of the Earth or another star in all directions, which would produce a kind of radiation analogous to the radiation of light; for in this way we would have a recession from the centre of aetherial matter, which would push down grosser bodies not having the same force of receding (as I will explain elsewhere) towards the centre, that is to say, would render them heavy.

(8) It remains that I should also consider what cause made the Earth's globe spherical when it was fluid; namely, the varied motion of the ambient fluid tending in every direction which (like the motion [232/233] that forms drops of oil in water) is disturbed by parts having a different nature, and so pushes the parts of the Earth down towards itself, while meanwhile the Earth itself pushes away from itself similarly heterogeneous parts, as I said a little while before.

(9) I am surprised how the most distinguished Objector induced himself to believe that by the hypothesis of infinite speed attributed to the circulating aether he could eliminate the difficulty that exercised the most distinguished Sturm and Bernoulli, among others. For heavy bodies are said to be pushed down towards the axis rather than the centre of the Earth not because the speed of the earthly circulation is greater at the equator than in the parallels, but because the aetherial matter itself, moved in smaller circles parallel to the equator and tending to recede from their centres, pushes the heavy bodies towards the centres of each circle, which centres do not fall on the centre of the Earth but on other points on its axis. So it is not the difference of velocity that is in question, but the difference of direction. Nor do I see how the defect in the latter hypothesis could be countered, unless we ascribe to the aether immediately producing gravity a magnetic motion, as it were, not in the equator and parallels, but in the meridians, as I already observed first in a published tract, then in a letter to Reverend Father Kochański, which proposed the same things.[9]

[8] *dispositionem*; this is very probably a mistake for *displosionem*, explosion, to which Gerhardt had silently corrected it. For Leibniz's account of his 'explosion theory of gravity', see Aiton (1972b, ch. VI) and Bertoloni Meli (1993, 166–8).

[9] Adam Adamandy Kochański (1631–1700) was a as a Polish Jesuit mathematician and natural philosopher with whom Leibniz had corresponded. Kochański found a famous approximation of π that is named after him, $\sqrt{\frac{40}{3} - 2\sqrt{3}}$, and was familiar with the elements of Leibniz's calculus.

(10) I do not have much time for the matter of the *first Cartesian element, and the globules of the second element*, creators of light; for I regard both of them as fictions, and hold for demonstrated that there is neither a first nor a second element in nature, and that light does not consist in what Descartes describes as an endeavour. Moreover, the same cause that forms drops of liquids also keeps the solar vortex in the world and the terrestrial one in the Sun's vortex, and draws its own limits around the exploded matter, so that it does not disperse as it tries to recede from the centre of the vortex.

(11) In every fluid, that is, the motion is varied in all directions, as we see when a little stick is given various motions in agitated water; moreover, this stick is perturbed when it is insufficiently pervious to the insertion of parts having a different texture and motion, and tries to repel obstacles and diminish them: although it is apparent that it resists those things less which are gathered into the most capacious figures having the same surface. Moreover, not just *vortices* ôr *bubbles* (with the nucleus removed), but also all things having their origin in consistency, that is to say, *cohesions*, and, so to speak, the warps of things and the bases of the textures proper to each mass, arise from a conspiring ratio of motion. This happens when, with these causes of original firmness established, bodies (that have some firmness already from this) come into greater or lesser contact, and thence, because of the resistance of the ambient bodies, cohesion can come to have arisen in ratios. [233/234] Motion, that is, or if you prefer, motive force, is the one thing that divides matter and renders it *heterogeneous*. From it come *congruity* and *incongruity*, since matter in itself is continuous and uniform, and indeed there is no other way that figures and real ôr actually determinate parts in it can be understood. Thus motion is also the principle of cohesion, and *fluidity* arises from varied motion, *firmness* from conspiring motion, as I already explained before; or rather nothing is so fluid that it does not have some firmness, nothing so firm that it does not have some degree of fluidity; but the denominations arise from their predominance in relation to sense.

(12) But now at last we must come to the defence of my own opinion, for which perhaps there may seem to be no very great need in the face of the objection, which is assumed in the controversy without touching on the force of my reasoning. Since, however, I have so much respect for the Objector that I think that what deceived him could also lead others astray, or rather because I acknowledge that my words fell somewhat short of the greatest clarity, I will try to see whether I can satisfy him as a fair judge. I will therefore demonstrate the proposition which he denies, and by taking the opportunity to set everything (I hope) in a clear light, I will finally reveal the source of the error.

But first there is an opportunity for excluding all quarrelling about words; for there will be those who allow themselves to define *force* by quantity of motion, and with the speed of a given body doubled, say that its force is doubled; and I do not deny this freedom to anyone, a freedom I ask to be conceded to me too. But since for us there is a real controversy, namely, over whether what is conserved is motion, or whether it is rather the same quantity of force in the sense accepted by me—that is, in the ratio compounded not of weight and speed, but weight and the height through which a body can be raised by an agent having the force—it would be easy for us to haggle about words. And so here I define there to be an *unequal force* when, if one of them is allowed to be substituted for the other, there could arise a perpetual mechanical motion; and the substituted one, in fact, is said to have a *greater force*, the other a *smaller one*; but if such an absurdity as perpetual motion cannot arise from the substitution of one for the other, we call their *forces equal*. With this definition supposed, the distinguished Objector will easily concede as a corollary that *the same force is conserved in bodies*, ôr that the power of the full cause is the same as that of the entire effect, or that the power of the preceding state is the same as that of the succeeding state arising from it; otherwise, if it were a bit stronger than the preceding one, perpetual mechanical motion there could arise from this. [234/235] It will not be denied that this kind of gain of force is impossible, I believe, so that in physics and mechanics reducing to a perpetual mechanical motion is therefore the same thing as reducing to absurdity. But in this same sense it is also necessary that *forces are in the compound ratio of weights and heights*, or, what is the same thing, that the sum of the products of the weights and the heights (to which the weights can be raised from those given) is conserved, but not the sum of the weights multiplied into the velocities, as the Cartesians persuade themselves. Which I will now demonstrate as follows:

Let us suppose (for the sake of example) a globe *A* of 4 pounds (Table VII, Fig. 4 [Fig. 13.2])[10] to descend from a height of one foot ${}_1AE$ along an inclined line ${}_1A_2A$, until it arrives on a horizontal plane *EF*, and that there it goes from ${}_2A$ to ${}_3A$ by the one degree of speed acquired through the descent. Next let there be at rest in the same horizontal plane another globe *B* of one pound, at the place ${}_1B$. Now let us suppose, next, that every power of the globe *A* must be transferred into the globe *B*, so that with *A* at rest in the place ${}_3A$ on the horizontal plane, only the globe *B* is then moving. We want to know how much speed must be received by the globe *B* in order for it to acquire just as much force as the globe *A* had. The Cartesians say that a globe *B* four times

[10] Fig. 4 is lacking the point *F*, which lies perpendicularly below ${}_3B$.

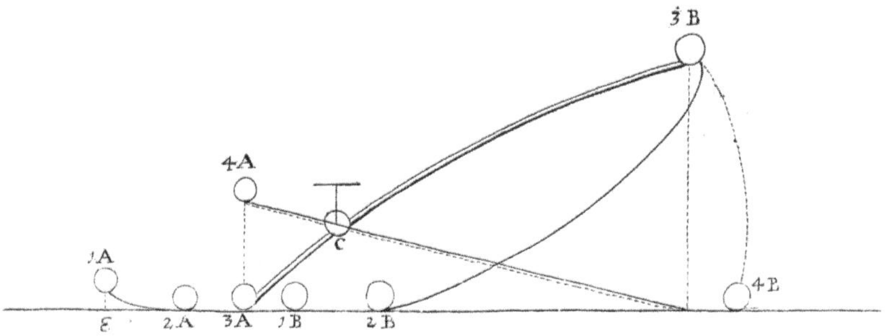

Fig. 13.2

smaller than *A* will receive a speed of 4 degrees, that is, four times greater than *A*'s speed; for *A* of 4 pounds with a speed of 1 will have just as much force as *B* of one pound with a speed of 4. But I will show that from such a substitution there can arise a perpetual motion, that is, an absurdity. For a body *B* of 1 pound having a speed of 4, by the aid of its motion directly upwards (as if by proceeding from $_1B$ to $_2B$ it runs onto the inclined line $_2B_3B$), could ascend towards $_3B$ ôr to a perpendicular height F_3B of 16 feet, since the body *A* had acquired a degree of speed of 1 by descending from a height of 1 foot, and so could again ascend to a height of 1 foot. Therefore it can ascend with a quadruple degree of speed to 16 feet; for the heights to which it can ascend by the force of the speeds are as the squares of the speeds. But now from this there will arise a perpetual motion, ôr an effect more powerful than the cause. For the globe *B* of one pound, when it is now raised to a height of 16 feet, could then be applied by us in such a way that by descending from there onto the horizontal plane at $_4B$ by a certain simple mechanism, for example, by means of an inclined rectilinear balance, it could raise the globe *A* of four pounds lying at the place $_3B$ on the horizontal plane, to a vertical height of nearly 4 feet. For let the balance reaching from $_3A$ to $_3B$ on a fulcrum or centre of equilibrium C, be divided [235/36] into arms of unequal length (although equal weight) C_3A and C_3B in such a way that the arm C_3B is a little more than four times the length of the arm C_3A. And so the globe *B*, falling onto the end of the balance $_3B$, overcomes and raises the globe *A* sitting at the other end of the balance, $_3A$; this is because the inverse ratio of the distances from the centre is greater (namely, four times greater, by construction) than the ratio of the weights (which, by hypothesis, is quadruple), and so *B*, by falling down onto the horizontal plane from $_3B$ to $_4B$ through the vertical height $_3BF$ of 16 feet, raises *A* from $_3A$ to $_4A$ to a vertical height of a little less than 4 feet, by a deficit as small as you like. In practice it suffices for *A* to be raised to a vertical height of around 3 feet, or even to some smaller height. Which is absurd. For at the beginning *A* was only raised to 1 foot, with *B* remaining on the horizontal; whereas now in the final state, with *B* restored to the horizontal, *A* is not restored to 1 foot, which the highest it could be with accidental impediments removed, but ascends to more than 3 and nearly 4 feet. And this by the force of its own descent from that one foot, albeit by the interposition of the body *B*, which, however, contributed no new force, but only the force we supposed *A* to have had. Thus we have gained almost triple the force, and have extracted it, so to speak, out of nothing: which is absurd, as indeed no intelligent person will dispute.

And now we will not be looking out for perpetual motion much longer. For it is easy to make the globe *A* return from $_4A$ to its original place, $_1A$, and before this in its journey (when it has fallen almost three feet) discharge some mechanical effect: to raise some other weights, set other machines turning, etc. And similarly in the meantime (while *A* is returning to $_1A$), if the place $_4B$ is assumed to be a little bit higher than the horizontal, the globe *B* could return by running down to the position $_1B$. And so everything would return to the original state, yet with an observable mechanical effect having been performed on top; and then the same game can be repeated. And these things can certainly not be permitted.

Hence, as I mention in passing, the inventors of perpetual motion will generally present it under the pretext of its plausibility as an occasion for new theorems, the fallacy in which is detected according to our present method by showing (by a certain kind of *Mechanical Algebra*, so to speak) a latent *equation between cause and effect*, which no art can violate. And for us this sort of equation has also been useful here for discovering the true laws of transference of motions. Thus it must be said that if *B* (of [236/37] one pound) should acquire the power of *A* (of four pounds), that is, if *A*, which alone was moving to begin with, should return to rest, with now a motion only existing in *B* which had previously been at rest; then, if every other agent or patient that would add a new force or absorb part of the previous one is removed, then *B* could only acquire double the speed that *A* had had. For in this way *B* could only ascend to four feet, and *A* (if you employ the balance) could only be restored to one foot, and in this way no such absurdity would arise as that the posterior state ôr effect would be stronger than the cause ôr the state from which it had its origin, and instead everything would be precisely balanced.

Thus, by the same calculation we have concluded against the Cartesians *that the quantity of motion must not always be conserved.* For since with *A* at first being in motion we would have a body of 4 endowed with a speed of 1;

whereas now after the transfer we would have a body 1 endowed with a speed of 2; so that the quantity of motion in the latter state of things would be only half that of the former. There will be other cases where the quantity of motion increases. And it is the same in most other transfers of force when bodies act on one another, that the *quantity of motion* differs from the *quantity of forces* as explained by us (which we have shown to be estimated by the *quantity of the effect*), and so cannot be conserved. And, in general, let body A be endowed with an initial speed e and body B with speed y, but after the interaction let body A have the speed (e) and body B the speed (y). And similarly let the heights to which the bodies A and B could ascend be (respectively) x and z before the interaction, and (x) and (z) after the interaction. Then I say that we should have $Ax + Bz = A(x) + B(z)$ in order for the power to be conserved. From which it certainly follows that it is not always possible for it to be the case that $Ae + By = A(e) + B(y)$, that is, that it is not possible for the same quantity of motion always to be conserved.

It remains for us to disclose the source of the error in a few words. And, indeed, the most distinguished Objector does not even need an infinite speed for the motive aether, and I very freely concede to him without proof that the speeds acquired or lost in free fall or free ascent are as the times. But I have shown the MOTIVE FORCES, that is, *those which are to be conserved*, should not be estimated by the degrees of speed. But a prejudice of the Schools has deceived most otherwise very learned men, in that they conceive motion and speed (degree of motion) as a certain real and absolute entity in things, and to be just like when the same quantity of salt is widely diffused in a smaller or greater expanse of water, which analogy Rohault also [237/238] used (as far as I remember).[11] Consequently it seems a wonder to them that the quantity of motion could be increased or diminished without the miracle of God's creating or annihilating it. But motion consists in a certain respect, so that, strictly speaking, it does not exist, any more than does time, or any other whole whose parts cannot be together at the same time, so it should be that much less of a wonder that the same quantity of it is not conserved. But motive force itself (ôr the state of things from which a change of place arises) is something absolute and subsistent, and to that extent its quantity is not provided for by nature. From which we also learn that *there is something else in things besides extension and motion*, which intelligent people know to be of great importance.

[11] Jacques Rohault (1618–1672) was a French natural philosopher and follower of Descartes, whose *Traité de physique* (Paris, 1671) became a standard text. It was later translated into English by Samuel Clarke.

For even if at first sight it seems that by doubling the speed of the same body its force is also doubled, this, however, cannot be admitted. But, you will say, if body *A* has some degree of speed, and a degree of speed equal to the first supervenes on the same body, what it had before certainly seems to be perfectly duplicated ôr repeated: but I reply by denying this, for the force will finally be exactly double what it was before when one adds to the body *A* having a speed of, say, 1, another body *B* equal to *A*, also having a speed of 1, and when this is done I agree that the force will be doubled too. How a multiplication of power is obtained without the body being multiplied, however, is achieved by the method explained above; by which considerations it was shown that the cause of their going astray was in the fact that they thought that power should be estimated not only from the effect that is produced, in which regard they were right, but also from the time in which it is produced, so that the power ought not to be estimated only by the compound ratio of weight and the height to which the weight can be raised by the potential. And it is certainly true that the ratio to the time should also be taken into account in producing those effects in which the same power can produce a greater effect if a longer time is allowed. This happens, for instance, when a globe having a given speed has the power of transferring its own weight in a horizontal plane through a given space in a given time. But the case is different in the effects and powers with which we are concerned here, where the force of acting is consumed, and whatever is endowed with force (for instance, a bow tensed to a certain degree, giving a body a certain velocity), if it expends all its action in lifting a certain body to a given height according to one mode of operation, no other mechanism or artifice will make the same weight rise higher, however much time is allowed. Whence a consideration of the time is to no avail. [238/39] For since that weight, falling down from this height, can (if we abstract away from the accidental impediments) precisely reproduce the very same force (as the said tension of that bow reproduces the said velocity of that body), it is certainly the case that if, after however much time, the weight is raised to a greater height, then by falling back down it could soon reproduce not only the force it was said to have before, but also bring about something else in addition, and would thus give a means of achieving a perpetual mechanical motion through a sufficient length of time. Which is absurd.

14

A Reply to the articles that the illustrious J. B. published in the May issue of these *Acta*[1]

G. W. Leibniz, *Acta eruditorum*, July 1690[2]

The illustrious gentleman has published two articles: one containing the solution to a problem that he once posed himself, the other occasioned by a problem of mine.[3] He presents a satisfying solution, I see, to his own problem once posed in the *Journal des Sçavans*.[4] But regarding my problem, of which the most famous Huygens had published the solution and I myself the synthesis, he offers an analysis following the laws of the new calculus that I published in these *Acta*, which I call differential or incremental calculus. Indeed, certainly not just anyone is up to the task of developing that analysis of my problem, especially since the art of this kind of calculus is still known only to a few, and I don't know anyone who has penetrated my reasoning better than this illustrious man. Moreover, he has proposed [358/359] to me another problem to be solved about which I will say something soon; before then, returning the favour, I will lay out the foundation of the solution that he himself published regarding his own problem in the aforementioned *Journal*.

[1] From the Latin: *Ad ea quae vir clarissimus J. B. mense majo in his actis publicavit responsio (166–172), Acta eruditorum*, July 1690, 358–60. (ESP 191–93; GM VI 193–203.) Translated by Jeffrey McDonough, Lea Schroeder, and Samuel Levey.

[2] J. B. is, of course, Jacob Bernoulli. As Leibniz relates, this article was prompted by the challenge laid down by Bernoulli in the May issue of the *Acta eruditorum*. There Bernoulli had provided a solution to the dice game problem he had set back in 1685, and effectively joined the circle of those testing new methods on outstanding mathematical problems by using the Leibnizian calculus to give his own solution to the problem of the isochronous curve, already solved by Huygens and Leibniz, and offering a new challenge problem of his own: the Galilean catenary problem (or '*chainette*'). Here Leibniz offers his own solutions to the dice game problem using infinite series expansions, and takes the notion of challenge problems a stage further by setting a deadline by which solutions to the catenary problem should be displayed.

[3] These articles of Bernoulli's, '*Analysis problematis*...' and '*Quaestiones nonnullæ de usuris*', were published consecutively in the May 1690 *Acta eruditorum* (pp. 217–23).

[4] '*Probleme proposé par M. Bernoulli*' (Bernoulli 1685).

Leibniz: Journal Articles on Natural Philosophy. Richard T. W. Arthur, Oxford University Press.
DOI: 10.1093/oso/9780192843531.003.0015

He puts it this way: Two players play with a die under the condition that the first player to throw a fixed number of points wins. To start, *A* begins with one throw and *B* with one throw; then *A* has two throws, and accordingly *B* has two throws; then *A* has three and *B* has three, and so on. *Or*: *A* begins with one throw, then *B* has two throws, then *A* three, thereafter *B* four, and so on until one of the two wins. The ratio of chances is sought. I present the case as follows:

Let $5{:}6 = n$, then $1{:}6 = 1 - n$.

In the former case,

1	n	n^2	n^3	n^4	n^5	n^6	n^7	n^8	n^9	n^{10}	n^{11}	etc.
A	*B*	*A*	*A*	*B*	*B*	*A*	*A*	*A*	*B*	*B*	*B*	etc.

The chance of *A* is,

$1 + n^2 + n^3 + n^6 + n^7 + n^8 + n^{12} + n^{14} + n^{15}$ etc., multiplied by $1 - n$,

which, after multiplication, yields

$1 - n^1 + n^2 - n^4 + n^6 - n^9 + n^{12} - n^{16}$ etc.

But the chance of *B* is,

$n + n^4 + n^5 + n^9 + n^{10} + n^{11} + n^{16} + n^{17} + n^{18} + n^{19}$ etc., multiplied by $1 - n$,

which, when the multiplication is actually done, yields

$n^1 - n^2 + n^4 - n^6 + n^9 - n^{12} + n^{16}$ etc.

In the latter case,

1	n	n^2	n^3	n^4	n^5	n^6	n^7	n^8	n^9	n^{10}	etc.
A	*B*	*B*	*A*	*A*	*A*	*B*	*B*	*B*	*B*	*A*	etc.

The chance of *A* is

$1 + n^3 + n^4 + n^5 + n^{10} + n^{11} + n^{12} + n^{13} + n^{14}$ etc., multiplied by $1 - n$,

ôr, with the multiplication done,

$1 - n^1 + n^3 - n^6 + n^{10} - n^{15}$ etc.

But the chance of *B* is,

$n + n^2 + n^6 + n^7 + n^8 + n^9 + n^{15} + n^{16} + n^{17} + n^{18} + n^{19} + n^{20}$ etc., multiplied by $1 - n$,

ôr, with the multiplication done,

$n^1 - n^3 + n^6 - n^{10} + n^{15} - n^{21}$ etc.

And in either case $A + B = 1$, assuming that unity is the whole right [*ius*] to the stake of the game. The same method succeeds in other similar cases even if [359/360] there are more players and dice, and hence a solution is easily obtained to any desired degree of precision. But what is especially pleasing about this problem is that although it appears quite simple, it nonetheless leads to series that have not yet been sufficiently examined.

He [Bernoulli] has proposed another problem for me in the following words: 'I now extend the author's analysis, established with the *differential calculus*, to this end: to invite him to offer an equal kind of service to the public and attempt the solution of the problem stated next, using his own method.' Moreover, he gives the problem itself as follows: 'To find the line into which a cord suspended by its two ends curves itself under its own weight.' Now, it is assumed that the cord remains of the same length, like a chain, and does not stretch or contract like a string. As far as I know, this proposed problem, famous since the time of Galileo, has not yet received a solution. I could therefore justly excuse myself from the charge assigned to me, especially since I am greatly distracted with other matters. But the courtesy of this illustrious man has made me loath to neglect his challenge. And so, I undertook it, which I had not tried even once before, and happily unlocked the closed avenues with my key.

This problem, however, is a little more involved than my earlier one, and will exhibit a certain unique use of our method; and so I thought it worthwhile to give others the opportunity to exercise their skill before the publication of my solution. For with this, as if with a touchstone, we will come to recognize the best methods; and that is what contributes most to the perfection of the science—especially since in this case there is no need of lengthy calculations, but only of skill. In particular, however, one should ask the most noble Mr Tschirnhaus, who promises brilliant things in this area (see the *Acta* from February of this year, pp. 68–69),[5] whether he is willing to test the strength of his method on this case as well. But if no one indicates that he has found a solution on his own by the end of the year, I will give mine, God willing. Similarly, it will not be unprofitable for the progress of science to make an attempt at the problem that I posed in place of the problem Huygens solved, viz: we seek 'the line which a heavy body running along uniformly recedes from a given point or approaches a given point'. Since problems of this kind are not in the power of Algebra or the common Specious, they will serve to stimulate those who grant too much to things they have learned, as if nothing further of great import remained to be investigated in this area, and who, not without detriment to the republic of letters, cease to work with the diligence necessary for science to grow.

[5] Tschirnhaus, *Methodus Curvas Determinandi, quæ formantur a radiis reflexis, quorum incidentes ut paralleli considerantur, per D. T.* ['A Method for Determining Curves which are formed by the reflecting of rays whose lines of incidence are considered parallel'] (Tschirnhaus 1690a).

15

An Opinion about the Motive Forces of Mechanics, offered by D. Papin against the objections of the most distinguished G. G. L.[1]

Denis Papin, *Acta eruditorum*, January 1691[2]

Recently, I have seen the reflections of the most illustrious *G. G. L.* in the *Acta eruditorum*, January 1690, p. 228, and I am very grateful to him for wanting to respond to my doubts about the way motive forces should be estimated; and also for seeing fit to propose new controversies against my little dissertation. To these I have various things I should reply, and that I would choose to reply; for I count myself honoured to contend with so celebrated an adversary. But I should beware that I do not abuse the patience of my readers. This is why concerning the new questions of our controversy that are less essential, it suffices for me to advise readers of this one thing, that the things most worth looking at in this matter will be found in the most acute Huygens' treatise, *On the Cause of Gravity* (which he recently published with Pierre van der Aa of Leiden in Holland).[3] There they could observe that the velocity of the matter bringing about gravity is much less than that which I had related in the *Acta eruditorum*. For since I had not taken part in researching this by setting up

[1] From the Latin: *Mechanicorum de Viribus Motricibus sententia, asserta a D. Papino adversus Cl. G. G. L. objectiones, Acta eruditorum*, January 1691, 6–13. Translated by Richard T. W. Arthur.

[2] Although Papin sent off this reply only five months after Leibniz's response of May 1690, it did not appear until January 1691. See O'Hara (forthcoming, ch. 3). In it Papin appeals to Huygens' *Discours* in defence of his own description of Huygens' views on gravity, and then returns to the controversy over whether the quantity of motive force is correctly estimated by the quantity of motion. Agreeing with Leibniz that a body has more power if it can produce a greater effect, Papin insists, however, that producing a greater effect is simply overcoming greater resistance. Turning to Leibniz's thought experiment, he notes that it depends on the assumption that one body can transfer its whole force to another, but denies the possibility of this. His argument continues with considerations concerning the difficulty that in the thought experiments considered, the tension of the lever must be taken into account.

[3] Christiaan Huygens, *Discours de la cause de la pesanteur*, in Huygens (1690).

Leibniz: Journal Articles on Natural Philosophy. Richard T. W. Arthur, Oxford University Press.
DOI: 10.1093/oso/9780192843531.003.0016

experiments, it easily happened that after a stretch of time my memory failed me. But I hope this lapse will be the more easily excused, in that even if the velocity is much less than I had believed, it is still so great that the consequences I deduced from it hardly departed from the truth. By means of the velocity of the said matter the author explains the acceleration of heavy bodies in descent; and then he shows (p. 148) that the same effect cannot be expected from a moderate velocity, and thus very evidently refutes the most distinguished L.'s argument taken from the tube (fig. 3). Next, by means of the same principle he demonstrates that equal degrees of speed accrue to a falling body in equal times; or, if you prefer, he shows that that effect must necessarily follow from the cause that is present. In the same book he discovers solutions to the difficulties raised by both the most distinguished L. and the most celebrated Bernoulli against the explanation of the cause of gravity by means of the centrifugal force of the circulating matter. And in this way, to readers who have not yet engaged with Huygens' book, I have easily indicated the business of how they can be satisfied concerning all the aforesaid difficulties.

It remains to come to our controversy: namely, whether the quantity of motive force is correctly estimated by the quantity of motion. And here I will adhere to the same method [6/7] which the most distinguished Leibniz himself followed: that is, I will give a definition; then from the definition I will demonstrate the opinion I am defending; and, finally, I will uncover the source of the error.

Here I observe first that: according to a notion we hold in common, the quantity of forces is estimated by the quantity of effects: for, it is asked, how much can something bring about? or how much power does it have? and this is after all the same thing: and the most distinguished Leibniz strongly urges that the quantity of power should be estimated in this way. In order to have a brief, clear definition that no one should reject, it must be given in this sense: *Of two bodies in motion, that one has more power which can produce a greater effect: if, on the other hand, neither of them is of this kind, then those bodies have equal forces.* It should be noted, however, that it is not the quantity of space traversed by the body that is a legitimate measure of the quantity of the effect, nor the quantity of time for which the motion continues; but rather the resistance that it overcomes. This is clear from the fact that in mechanics it is held that spaces however great can be traversed horizontally, and that a horizontal motion can continue for ages without any diminution of power. This is because we suppose that no cause is needed for the continuing of a horizontal motion, that is, no resistance has to be overcome. It is established, then, that in the definition supplied above, instead of the phrase *what can produce a greater effect*, this other phrase can be safely substituted, *what can overcome a greater resistance*. For to produce an effect and to overcome a resistance are one and the same. On the other hand, the remaining definitions from the height of the vertical ascent, or from the duration of the time of ascent, or from the impossibility of deriving a perpetual motion, cannot be accepted, unless it is first established that those things in fact mean the same thing as this definition we have given, or entail it; and therefore, other things being equal, this definition deserves to be supposed beforehand. It will be very easy for me to demonstrate from it that the quantity of force is to be estimated by the quantity of motion, in accordance with Descartes's intention; and also to do this by employing the very example the most distinguished Leibniz proposed in order to prove the contrary.

Let us suppose that a globe *A* of 4 pounds (fig. 1 [Fig. 15.1]) descends from a height AE along an inclined plane ${}_1A_2A$ until it reaches the horizontal plane EF; and that there it runs from ${}_2A$ to ${}_3A$ by one degree of speed acquired through its descent. Next, in the same horizontal plane another globe *B* of one pound is at rest at place ${}_1B$. Let us suppose further that all the power of globe *A* must be transferred to the globe *B*, so that, with A at rest in the place ${}_3A$ on [7/8] the horizontal, only then does the globe *B* move. Then it is asked how much speed the globe *B* should receive in order for it to receive as much force as the globe *A* has. I reply with the Cartesians and almost all the other mechanical philosophers that a globe *B* four times smaller than *A* will accept four degrees of speed, that is, a speed four times greater than *A*'s; for in this way the powers on each side are equal. I prove it as follows: it is known that the body *B* should ascend with 4 degrees of speed, and overcome each resistance of gravity it comes up against through 4 times: if the body *A* with one degree of speed and on a similarly inclined plane were to be able to ascend through 1 time, it is moreover certain that the body *B* within 4 times will come up against the same quantity of resistance as the body *A* within 1 time, on similarly

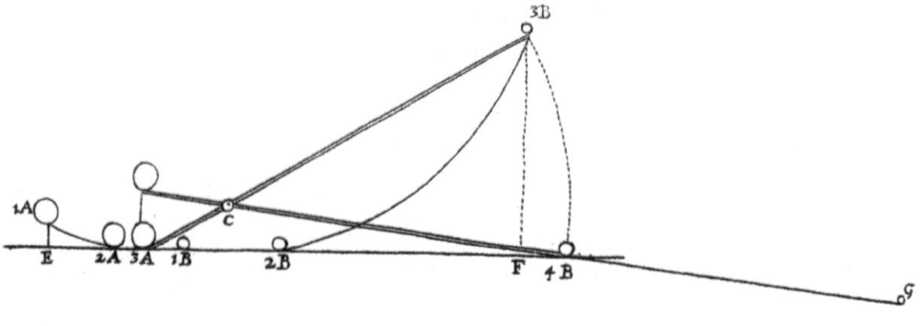

Fig. 15.1

inclined planes; therefore, since with the above degrees of speed neither of these bodies could overcome more resistance than the other, their powers will be equal, by the definition posited above. If the minor premise is denied to me, it is very easily proved. For it is established that the quantity of resistance here does not arise from the quantity of space traversed, but from the impressions of gravity pressing it downwards. But since the velocity of those impressions is judged to be infinite, it evidently follows that the impressions on the body *B* through 4 times will be four times more numerous than on the body *A* through one time; and conversely, the impressions on the body *A* of 4 pounds (provided this occurs on the same inclined plane) will be four times stronger than on the body *B* of one pound; since, therefore, the number of the former precisely compensates for the strength of the latter, it follows that an equal resistance is overcome in both cases, and that therefore the powers are equal in both cases, QED. Hence it is easily seen that if (according to the distinguished Leibniz's definition) there a smaller speed is transferred into the body *B*, it would not be able to overcome so much resistance: and, therefore, it would not have a power equal to the power of *A*.

If the distinguished Leibniz judges that there is something to be objected to in this discussion, I ask that he deign to show explicitly which of my propositions is false, or which conclusion has not been legitimately inferred. For in this way we will proceed to the end of the controversy correctly, and without ambiguities, and not waste the time of our Readers. Moreover, lest he reckon that I am proposing laws other than those he himself complies with, I ask that he should attend to how I responded, in the April 1689 issue, to the objections that he had proposed in the *Acta eruditorum* of May 1686. His objection was urged against this hypothesis, that [8/9] *the spaces traversed in ascending are proportional to the motive forces.* But I explicitly denied that proposition, and gave a reason for denying it: therefore, he will see whether he spoke fairly when he said on p. 228 of the 1690 *Acta* that I had not sufficiently touched on the state of the controversy: in fact, nothing more could have been responded on the matter. Now I will have enough trouble touching on his new argumentation no less strictly, and detecting the error in it.

The most distinguished gentleman proposes on p. 235 of the 1690 *Acta* the very example I brought forward above, and asserts that the annexed response of the Cartesians is absurd. This is because there follows from it a perpetual motion, which all intelligent people admit is absurd. But how by means of a transference of the whole power of body *A* into body *B*, in accordance with Descartes, a perpetual motion could be obtained, he demonstrates very clearly; and so in this way the Cartesians are reduced to absurdity. Now, I also admit that perpetual motion is absurd, and that the distinguished gentleman's demonstration by the supposed transference is legitimate; but I completely deny the possibility of this hypothesis of the transference of the whole power of body *A* into body *B*. It is obvious that if the said transference could not happen anywhere in nature or for any reason, then the perpetual motion that was expected from it would also remain impossible, and the Cartesians would not be reduced to this absurdity.

I could stop here. For thus far the opinion which is underpinned by the most valid demonstrations of the mechanical philosophers, remains untroubled by the designs of adversaries. And it is incumbent on Mr Leibniz to prove either that the translation which I deny in nature is possible, or at least that its possibility follows from the doctrines of his adversaries. For me it is enough to promise here publicly that if he indicates to me some reason whereby the whole motive force, without a miracle, can be transferred from a larger body into a smaller one at rest, then I will either grant perpetual motion or hold up my hands in defeat. In fact, however, in order for him to be able to oppose me more appropriately and securely, I judge it would not be beside the point if by some specimen I showed how I am prepared to defend myself.

First: suppose he says that the whole motive force of a body *A* of 4 pounds (fig. 2 [Fig 15.2]) can be transferred into a body *B* of 1 pound, when it is applied to the lever *CB* movable around the axis at *C*: for if the body *A* strikes the point *E* and the distance *CB* is four times the distance *CE*, then the body of 1 pound at *B* ought to offer the same force as the body of 4 pounds at the end of the lever *E*. But the body of 4 [9/10] pounds at *E* could receive the whole force of the body *A*. Therefore, the body of 1 pound at *B* ought to receive the same whole force. I respond by denying that the body of 1 pound at *B* ought to offer the same force as the body of 4 pounds at *E*. For while the body of 1 pound is at rest at *B*, and the body *A* impinges on the point *E*, the power of the body *A* does not act on body *B* alone. For it also exerts itself on the fulcrum C which has more effect the less the distance *CE* is than the distance *CB*, according to the very well known rule of mechanics. What if someone insists that the fulcrum is immobile, and thus receives no motion or force from the body *A*? I reply that nothing is immobile except with respect to the senses. For Huygens' rule that *A body carried by a motion however small and slow, if it collides with another body at rest of however small a bulk, impresses some motion on it*, is most certain, contrary to Descartes. By this artifice, therefore, the required transference of all of the power of body *A* into body *B* is by no means obtained.

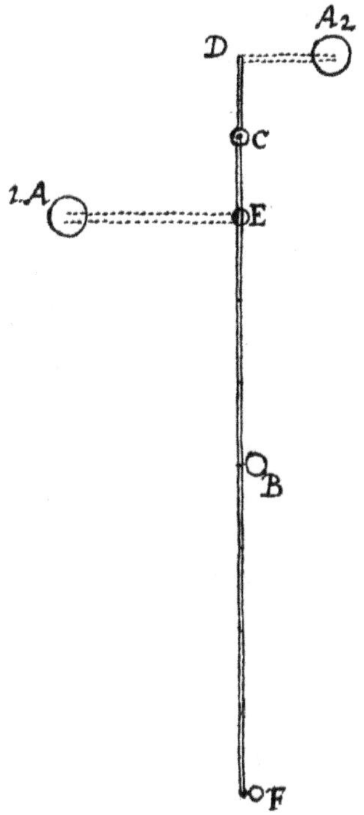

Fig. 15.2

Now suppose we employ a lever of another kind to bring it about that body *A* hits the lever at *D* from another part of the pivot, along the direction $_2AD$. I reply: this body *B* is impelled in the opposite direction by the reaction of the pivot *C*. More of this reaction is received at the body *A* the smaller the distance *DC* is than the distance *BC*. So this artifice does not succeed any more happily than the previous one.

Still another argument could be proposed, according to which it seems that a perpetual motion could be brought about according to the Cartesians, notwithstanding the impression that the pivot receives. For let the lever *CB* be produced to *F*, so that *FC* is twice as long as *BC*, and then let the distant body *B* of 1 pound be substituted by one of ½ pound at *F*. It is certain that ½ pound at *F* has as much force as one pound at *B*, and so the same impression will always be made on the fulcrum *C*. Then nothing prevents the remaining force that was to be transferred into the body *B* from now transferring into the body *F*. And thus, by producing the lever further, and diminishing the motion of the body in the same ratio, we will have the same power being transferred into the smaller body, and always a smaller body at will. If, then, power and motion are one and the same, in smaller bodies the velocity should increase in the same ratio as their bulk decreases. The height of ascent will therefore increase as the square of the speed; and so the said bodies could very easily be brought to such a height that, notwithstanding the decrease in bulk, the possibility of a perpetual motion would follow, according to [10/11] the demonstration provided by the most distinguished Leibniz on p. 235. To this I reply that the validity of this argument cannot be denied if we suppose the lever perfectly hard and rigid, so that the same motive power is transferred just as easily to the distant parts as it is to the near ones. But, in fact, there is no perfect hardness of this kind in the nature of things; and levers made of whatever kind of matter, whether supporting a hanging weight (as Mr Mariotte has explained in his treatise *On the movement of waters*, p. 373), or overcoming some other resistance, necessarily suffer some degree of tension proportional to the overcoming of resistance. Now the longer the lever is, the more parts there must be for reducing that tension, and for conserving it, as long as the action lasts. For example, if the resistance is at *B*, only the length *CB* will have to be supported; if, on the other hand, the resistance is at *F*, force will have to be expended not only for sustaining the part *CB*, but also the part *BF*. And the longer the action lasts, the more power is required for overcoming the continual strivings of those tensed parts which endeavour to restore themselves by their elasticity. Therefore, it is obvious that if instead one substitutes for the body *B* another smaller one such as *F* that is more distant from the centre *C*, that body will necessarily receive a part of the power of the body *A* that is smaller in proportion as the distance is greater. For the remainder of the power is expended in bringing about and conserving the above-mentioned tension. So, it is false that the same quantity of power can be transferred into smaller and smaller bodies. The matter could be further clarified by introducing a hypothesis according to which the tension and elasticity are clearly and mechanically explained. But, in fact, this is not the place to take up such things, not only for the sake of brevity, but also because for me it is enough to have shown a deficiency in the most distinguished Leibniz's argument. For since the demonstration given above proving the opinion of the Cartesians is most evident, then, as long as their adversaries' opinion does not enjoy the same certainty, it is clear enough that the former should be preferred.

Since perhaps it will seem wonderful to many people that no one that I know has up till now observed this last resistance sought from the tension of the lever, I will here briefly suggest the reason why. On the balance depicted in fig.1 [Fig. 15.1], if we hang for example a weight of 4 pounds at a distance of one inch from the centre; and from another part we annex *B* of 1 pound at a distance of 4 inches, we will have equilibrium, and at once the two arms will be reduced to the tension required for supporting the two weights, and thus [11/12] they will remain in tension, since the impressions of gravity on the weights will always be continued. Now if we add one ounce to the 4 pound weight, that will immediately be pushed down. For although the tension in the lever should be increased in order to impress motion on the opposite weight, since that motion is slower, so tiny an increase in tension is required that a weight of one ounce is enough to bring it about. If, however, instead of adding one pound at a distance of 4 inches we were to add half a pound at a distance of 8 inches at *G*, it would also make an equilibrium with 4 pounds at a distance of 1 inch. And one ounce added to the same 4 pounds would be enough to push it down, but with a slower motion than when one pound remains at a distance of 4 inches. For since the longer part of the lever is to be supported by smaller appropriate weight, the motive power of resistance opposes the resistance too much unless the slower motion in turn diminishes the resistance. There is therefore actually some difference in speed in the two cases here; but since that difference is very insignificant, it would hardly ever be observed, and it is commonly held for certain that whether we hang 1 pound at a distance of 4 inches or ½ pound at a distance of 8 inches, absolutely the same effect will follow. In fact, it happens that when the lever is under tension for a long time, and therefore resists more, it will likewise slow down the descent of the weight more, and will receive so many more impressions of gravity, which suffices for an increase in resistance. The result is that the ascending weight attached to the longer arm reaches the same height at the end of its ascent as if no resistance proceeded from the tension of the lever, which is therefore of no importance in those machines. But the effects of those machines are subsumed under a very exact measure, from which the error is made more evident.

Nevertheless, the said maximum resistance is to be observed in machines which principally require velocity, as may be seen in the examination of the spouting machine by Mr Perrault (*Acta*, April 1689). For there he considered a lever as perfectly hard, and paid no attention to the tension by which it acts, and so the effects ascribed to the said machine were found to be much greater in practice. This is because the tension, being conserved through the whole length of the lever *CD* (see fig. 3 [Fig. 15.3]), very much retarded the descent of the weight *A*, as a result of which also the velocity of the globe *D* would be diminished, and the height of its ascent even more so. In the same examination it could also be objected that I had not paid attention to the gravity by which the globe *D* strives downwards, and detracts from the efficacy of the weight [12/13] *A*, to the extent that if the globe were increased to 200 pounds, the weight *A* would absolutely sustain it in equilibrium. But, in fact,

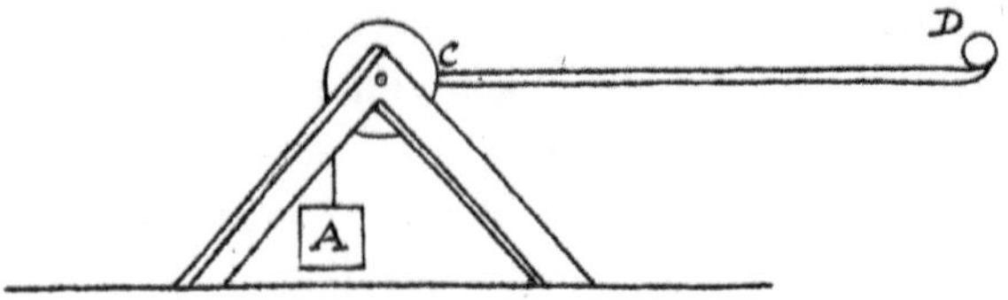

Fig. 15.3

I have estimated that here the lever can be considered as if it were moving horizontally: for a mechanism could easily be arranged for this affair in such a way that by its horizontal motion the lever would impress a speed on the globe, which could then be directed to a destined purpose by meeting some inclined plane. But it is clear that in such an arrangement the globe would not resist the weight *A* by any gravity, nor would it ever make an equilibrium, but solely by its own mass positioned far from the centre, would increase the motion of the mass by the moment of gravity by which the weight *A* is pressed down. Moreover, for the sake of brevity and ease, I neglected to make mention of these things there, because I had sufficiently achieved my intention. For what I had in mind was to compare with each other the effects which that machine could produce if it were set in motion by evacuated tubes, and those it could produce if it were set in motion by common weights, so that from this the effectiveness of evacuated tubes would be brought to light. But that effectiveness was apparent precisely as if nothing was omitted; because, in fact, in both arrangements the same omission was made, and thus there always resulted the same reason in each case for the effects. Nonetheless, however, given this opportunity, I considered it not imprudent to point out these things briefly: especially since I understood that those omissions could be made good by the readers by means of the letters which the illustrious Huygens recently saw fit to give me.

In the *Nouvelles de la République des Lettres,* A, 1687, p. 141 & ff., there are some difficulties which the most distinguished Leibniz himself proposes, and which he admits follow necessarily from his own doctrine:[4] but whether the things he offers there are equally clear and intelligible as those I propose here, the readers may judge.

[4] This refers to Leibniz's reply to Catelan, *Nouvelles de la République des Lettres*, February 1687 (text [5]). The page in question is where Leibniz criticizes the third of Malebranche's Rules of Collision, and numbers 3, 4, 6, and 7 of Descartes's Rules. But the difficulties that Leibniz enumerates are difficulties for Malebranche's and Descartes's doctrines, not his own.

16

Addition to the Sketch on the Resistance of the Medium published in these *Acta* in February 1689[1]

G. W. Leibniz, *Acta eruditorum*, April 1691[2]

After I published certain thoughts on the resistance of the medium in these *Acta*, there came into my hands what Huygens and Newton, preeminent in their mathematical knowledge of nature, have commented on the same subject in their newest works. I noticed, however, that they had touched only on (what I call) respective resistance, that is, the kind a body experiences in a liquid lacking any observable tenacity, such as air; but not absolute resistance, which arises from the tenacity of the medium, or from the roughness of the surface of contact bringing about friction. I have already shown that there is a great difference between these two kinds of resistance, since respective resistance occurs with respect to the speed of the moving body, and increases when the speed is increased, while absolute resistance does not do likewise. Concerning respective resistance, we have built on the same foundations, even if at first sight it might seem otherwise. For they state that the resistances are as the squares of the velocities, whereas I have said that absolutely speaking the resistances (which I estimate from the decrements in velocity arising from the density of the medium) stand in a compound ratio [177/178] of the velocities and elements of the spaces that they have begun to traverse with the corresponding velocities; hence now when the elements of time are assumed equal (in which case the elements of space to be traversed are proportional to the velocities), the resistances will evidently be as the squares

[1] From the Latin: '*O. V. E. Additio ad Schediasmam de Medii Resistentia publicatum in Actis mensis Febr. 1689*', *Acta eruditorum*, April 1691, 177–8. (The *Schediasma* was actually published in January 1689, not February.) (ESP 194–95; GM VI 143–44.) Translated by Tzuchien Tho and Richard T. W. Arthur.

[2] In this article Leibniz corrects a mistake in his *Schediasma* that was pointed out to him by Huygens in their correspondence. Leibniz also uses the occasion to make some comments on the relationship between his treatment of motion in a resisting medium and those of Huygens and Newton.

Leibniz: Journal Articles on Natural Philosophy. Richard T. W. Arthur, Oxford University Press. © Richard T. W. Arthur, Richard Francks, Samuel Levey, Jeffrey K. McDonough, Lea Aurelia Schroeder, and Tzuchien Tho 2023.
DOI: 10.1093/oso/9780192843531.003.0017

of the velocities; which I had also noted under art. 5 prop. 3. Nor does the conclusion differ about the relation between the times and velocities in the case of a heavy body descending through a medium. For this, Newton reduced it to a hyperbolic sector, while Huygens reduced it to an infinite series which he found to depend on the quadrature of the hyperbola, and I reduced it to logarithms in art. 5 prop. 4 as offering the most perfect way of expressing such things.

Specifically, let the *maximum* velocity be *a*, the present velocity be [*v*],[3] the time *t*, and we get

$$t = \int dv \cdot aa : (aa - vv)$$

With this supposed, the *t* are as the logarithms of the ratios of $a + v$ to $a - v$;[4] so we also get:[5]

$$t = \frac{1}{1}v + \frac{1}{3}v^3 + \frac{1}{5}v^4 + \frac{1}{7}v^7 \ldots$$

putting *a* as 1.

Regarding the composition of motion in a resisting medium, the most celebrated Huygens quite correctly advised that in that case it does not take place simply, as it does in free motion.[6] So what I expounded in articles 3 and 6 should be understood in such a way that, for example, it is as if some body is moving in a medium according to a single law of compound motion, and in this very body (for instance, a ship) is contained a medium of the same nature as the previous one in which another body is carried in turn, whose motion is now composed of the motion in common with the ship and its own proper motion, just as if making a projectile motion, thus behaving as we have described.[7]

[3] Here '*v*' was misprinted as '*b*' in the *Acta eruditorum*.

[4] That is, $dv/(a^2 - v^2) = (1/2a)\,dv\,(a - v + a + v)/(a^2 - v^2) = (1/2a)dv\{[(1/(a + v)] + [(1/(a - v)]$, which integrates to $\ln(a + v) - \ln(a - v) = \ln[(a + v)/(a - v)]$, if we neglect the constant factor $(1/2a)$ and the constant of integration.

[5] Here Leibniz has expanded $1/(1 - v^2)$ as a power series, $1 + v^2 + v^4 + v^6 + \ldots$, and integrated term-wise.

[6] Huygens had discovered in his *Discours* that the composition of motions could not be effected in the usual way if resistance is proportional to the square of the speed, for then the components of resistance are not proportional to the squares of the components of the speed. He judged that 'it is extremely difficult, if not impossible, to solve this problem' *Discours*, p. 175; but Johann Bernoulli did solve it, using Leibniz's calculus, in his article in the *Acta eruditorum*, of May 1719, 216.

[7] As Aiton (1972a) has noted, the mistake only occurs in article 6 of the *Schediasma*, not article 3. Indeed, in acknowledging the mistake in his letter to Huygens of 2 March 1691, Leibniz only mentions his article 6 in this connection, 'where the composition of the two motions cannot take place, and in order for my article 6 to hold one would need a particular hypothesis' (OCH, x, 50).

17

On the Line into which a flexible body curves itself under its own weight, and its remarkable usefulness for discovering any number of mean proportionals and logarithms[1]

G. W. Leibniz, *Acta eruditorum*, June 1691[2]

The problem of the *Catenary or Funicular Line* has two uses: one is to promote the art of discovery ôr analysis, which thus far has not been extended enough to such matters; the other is to improve the practice of construction. Indeed, I have found this line as useful in its application as it is easy to produce, nor is it second to any of the transcendental lines. For it can be described and prepared without difficulty *by a kind of physical construction* by the suspension of a string, or rather a chain (whose length does not change). And once the line has been described, mean proportionals or logarithms as well as the quadrature of the hyperbola can be produced by means of it. Galileo was the first to consider this line, but he did not grasp its nature, for it is not a parabola as he

[1] From the Latin: *De linea in quam flexile se pondere proprio curvat, ejusque usu insigni ad inveniendas quotcunque medias proportionales & Logorithmos*, June 1691, 277–81. (ESP 206–11; GM V 243–47.) Translated by Jeffrey McDonough, Lea Schroeder, and Samuel Levey.

[2] As Leibniz explains, the problem of determining the shape and properties of the curve made by a hanging chain, the catenary (or *chainette* in French) was an as yet unsolved one at the time, and one that he had been asked to solve by Jacob Bernoulli. Having solved it, he had set it up as a challenge problem for mathematicians of the time to solve within a year. Here he gives his solution (its publication delayed by some months as a result of his long journey to Italy). Leibniz's pride at having solved it is evident. First he gives its geometric construction by reference to the logarithmic curve, allowing logarithms to be calculated from the physically constructed curve, then shows how to determine the tangent, rectification of the curve (arc length), quadrature (area under the curve), centre of gravity of any arc of the curve (including the (variational) result that the centre of gravity of the catenary curve is the lowest possible for such a curve suspended between two points), centre of gravity of an area bounded by the curve, surfaces and volumes generated by its rotation, and the infinite series for $\sinh^{-1}a$ (or, arcsinh a). All these results are given without accompanying demonstrations; we have included sketches of such demonstrations in the notes.

Leibniz: Journal Articles on Natural Philosophy. Richard T. W. Arthur, Oxford University Press.
DOI: 10.1093/oso/9780192843531.003.0018

had suspected. Joachim Jungius, an eminent philosopher and mathematician of our century, who long before Descartes had many splendid thoughts concerning the improvement of the sciences, excluded the parabola by performing some initial calculations and experiments, but he did not replace it with the true line. Since then, the problem has been tried by many but solved by no one, until recently a most learned mathematician provided me with the occasion for treating it. For after the illustrious Bernoulli had successfully applied to certain problems my Analysis of Infinites, which is expressed by means of the differential calculus introduced on my recommendation in the May issue of last year's *Acta* (p. 218 ff.), he publicly asked me to test whether our kind of calculus could also be extended to problems of this kind, such as the discovery of the catenary line.[3]

When I took up the challenge for his sake, not only did I succeed—and, unless I am mistaken, I am the first to have solved this famous problem—but I also discovered the line to have extraordinary uses. That is the reason why, following the example of Blaise Pascal and others, I invited mathematicians to undertake the same inquiry within a certain period of time determined in advance: for the sake of testing their methods, in order to see what those people might have to offer who would apply methods that are perhaps different from the one that Bernoulli and I use. Only two mathematicians indicated that they had succeeded before the time had elapsed: Christiaan Huygens, whose great merit in the literature is known to all; and—together with his ingenious and very learned younger brother—Bernoulli himself, whose publications make us hope for further [277/278] splendid things from each of them. I think he has really proven what I had indicated, namely, that our way of calculating extends also to this problem and now allows us access to things that were previously considered to be very difficult. But I propose to set out those things that I have discovered; a comparison[4] will show what the others have achieved.

Let the line be constructed geometrically without the aid of a string or chain and without supposition of quadratures, by a construction of the kind that, in my opinion, is the most perfect and most suitable for transcendentals. Let there be any two straight lines having a certain determined and invariable ratio

[3] Jacob Bernoulli, *Analysis problematis antehac propositi, . . .*, (Bernoulli May 1690a), making his request for Leibniz to apply his calculus to the catenary on p. 219: 'This in turn is the problem proposed: *To find what curve a slack cord makes when freely suspended between two fixed points.* I assume, moreover, that the cord is a line maximally flexible in all its parts.'

[4] Here Leibniz refers to the article he will publish in the *Acta* of September 1691 ([18] below), in which he describes the solutions of Huygens and Bernoulli and compares them with his own.

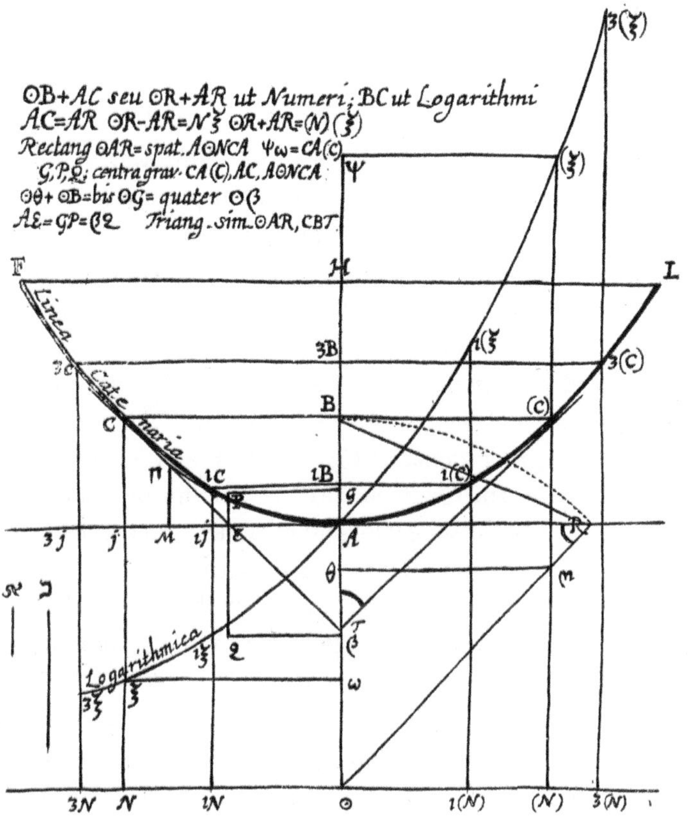

Fig. 17.1

with each other, such as that between א and ב represented here. Once this ratio is known, the rest follows by ordinary geometry.

Let the indefinite straight line $\odot N$ be parallel to the horizontal, and let $\odot A$ be perpendicular to it, equal to $\odot_3 N$; and let ${}_3N_3\xi$ be a vertical line above ${}_3N$ that is to $\odot$A as א is to ב.[5] [Fig. 17.1] The mean proportional ${}_1N_1\xi$ between $\odot A$ and ${}_3N_3\xi$ is sought, and likewise that between ${}_1N_1\xi$ and ${}_3N_3\xi$; and in this way, by further seeking mean proportionals and—once these are found—third proportionals, the line $\xi\xi A(\xi)(\xi)$ will be described and extended. This line will be of such a nature that if the intervals—for example, ${}_3N_3\xi$ ${}_3N_1N$, ${}_1N\odot$, $\odot_1(N)$, ${}_1(N)_3(N)$ etc.—are assumed to be equal, the ordinates ${}_3N_3\xi$, ${}_1N_1\xi$, $\odot A$,

[5] Here the ratio of א to ב acts as a parameter for the curve. The abscissas ${}_3N_3\xi$, ${}_3N_1N$, etc. (which we may designate x) from the base line to the curve are drawn vertically, the ordinates $\odot_1 N$, $\odot_3 N$, etc. (which we may designate y) are horizontal. If $y = ln\ x$, then when the x are in geometric progression, so that $x_2 = x_1^2$, etc. the y will be in arithmetic progression, $y_2 = 2y_1$, etc.

${}_1(N)_1\xi$, ${}_3(N)_3(\xi)$ are in a continual geometric progression. I am accustomed to call such a line Logarithmic. Now, assuming that $\odot N$ and $\odot$ (N) are equal, NC or $(N)(C)$—equal to half the sum of $N\xi$ and $(N)(\xi)$—is raised above N or (N), and C or (C) will be a point of *the catenary line* $FCA(C)L$, of which as many points as you like can be assigned geometrically in this way.[6]

Conversely, if the catenary line is constructed physically by the aid of a hanging string or chain, as many mean proportionals as you like can be exhibited with its help, and the logarithm of any given number or the number of any given logarithm can be found. In this way, if the logarithm of the number $\odot\omega$ is sought, assuming that the logarithm of $\odot A$ (I take $\odot A$ as unity, and shall also call it the *parameter*) is equal to zero; or, what amounts to the same, if the logarithm of the ratio between $\odot A$ and $\odot\omega$ is sought, let $\odot\psi$ be taken as the third proportional of $\odot\omega$ and $\odot A$, and let $\odot B$, the abscissa, be half the sum of $\odot\omega$ and $\odot\psi$; then the corresponding ordinate BC or $\odot N$ of the catenary line will be the sought logarithm of the given number. Conversely, with the logarithm $\odot N$ given, and with the double of the vertical NC drawn to the catenary, one must cut it into two parts such that the mean proportional between the segments is equal [278/279] to the given $\odot A$ (or unity) (which is very easy). The two segments *will correspond to the sought numbers of the given logarithm*, one greater than, the other smaller than unity. Alternatively: having found NC ôr $\odot R$ as was stated (by taking the point R on the horizontal AR in such a way that we have $\odot R$ equal to $\odot B$ or NC), the sum and the difference of the straight lines $\odot R$ and AR will be the two numbers corresponding to the given logarithm, one greater than, one less than one. For the difference of $\odot R$ and AR is $N\xi$, and their sum is $(N)(\xi)$; as in turn $\odot R$ is the semi-sum, and AR the semi-difference of $(N)(\xi)$ and $N\xi$.[7]

There follow *solutions for the primary problems* that are usually proposed about lines.

To draw the tangent at a given point C on the line: On the horizontal AR through the vertex A, let R be taken in such a way that $\odot R$ is equal to the given $\odot B$, and CT, drawn antiparallel to $\odot R$ (meeting the axis $A\odot$ at T) will be the

[6] If we let $z = NC = (N)(C) = ½\,[N\xi + (N)(\xi)]$ then in modern terms, since $x = e^y$, this is $½\,[x(y) + x(-y)] = ½\,[e^y + e^{-y}] = \cosh y$. So the catenary curve has the form of a hyperbolic cosine.

[7] Here R plays the role of a determinant. Since $\odot R^2 = \odot A^2 + AR^2$, $\odot A = 1$, and $\odot R = \cosh y$, we have $AR^2 = \odot R^2 - 1 = \cosh^2 y - 1 = \sinh^2 y$. Thus $AR = \sinh y$. Therefore $\odot R - AR = \cosh y - \sinh y = e^{-y}$, and $\odot R + AR = \cosh y + \sinh y = e^y$. So y is the logarithm of $\odot R + AR$. Geometrically, if NC and $(N)(C)$ are taken on either side of the axis under the catenary curve, both equal to $\odot B$, the shorter segment $N\xi$ under the logarithmic curve on one side of the axis is the difference between $\odot R$ and AR, while the longer segment $(N)(\xi)$ under the logarithmic curve on the other side of the axis is their sum, where $\odot R$ is constructed as the arithmetic mean of successive values of the abscissa, with AR their geometric mean.

tangent sought. For the sake of brevity, here I call ⊙*R* and *TC antiparallel* if the angles *AR*⊙ and *BCT* that they make with the parallels *AR* and *BC* are not equal, but complementary. And the right triangles ⊙*AR* and *CBT* are similar.[8]

To find the straight line equal to an arc of the catenary: describe a circle with a centre ⊙ and a radius ⊙*B* that cuts the horizontal line running through *A* at *R*; *AR* will be equal to the given arc *AC*. From what has been said above, it is also clear that $\psi\omega$ will be equal to the catenary *CA*(*C*). If the catenary *CA*(*C*) were equal to twice the parameter (i.e. if *AC* or *AR* were equal to ⊙*A*), the inclination of the catenary to the horizontal at *C* (i.e. the angle *BCT*) would be 45 degrees, and therefore the angle *CT*(*C*) would be a right angle.[9]

To find the quadrature of the space enclosed by the catenary line and one or more straight lines: with the point *R* found as before, the rectangle ⊙*AR* will be equal to the quadrilineum *A*⊙*NCA*. From here on it is easy to find any other partial area. It is also clear that the arcs are proportional to the areas of the quadrilinea.[10]

To find the centre of gravity of the catenary or any part of it. After finding the fourth proportional ⊙θ to the arc *AC* or *AR*, of the ordinate *BC*, and the parameter ⊙*A*, let the abscissa ⊙*B* be added; then the half-sum of this, ⊙*G*, will give the centre of gravity of the catenary *CA*(*C*).[11] Furthermore, let the tangent *CT* cut the horizontal through *A* at *E*, and let the rectangle *GAEP* be completed, then *P* will be the centre of gravity of the arc *AC*. The distance of the centre of gravity of any other arc like C_1C from the axis is *AM*, assuming πM is perpendicular to the horizontal running through the vertex, taken from the intersection point π of the tangents $C\pi$ and ${}_1C\pi$; although its centre of gravity is also easily obtained from the centre of gravity of the arcs *AC*, A_1C. [279/280] Hence *BG* is also obtained, the greatest possible descent of the centre of a cord or chain or any other flexible, non-extendible string [*lineae*] suspended at the two extremities *C* and (*C*), with a given length $\psi\omega$; for if it

[8] In modern terms, if $x = \cosh y$, then $dx/dy = \sinh y$.

[9] This demonstrates the remarkable result that the catenary curve—a transcendental curve—is rectifiable. An analysis of the conditions for the catenary yields the differential equation for the curve $zdy = adx$, where a is a constant, here set equal to 1, and z the length along the curve. Since $x = \cosh y$, we have $dx = \sinh y dy$. The arc length of an element of the curve is $dz = \sqrt{(dx^2 + dy^2)} = \sqrt{(1 + \sinh^2 y)}dy = \cosh y\, dy$, so that $z = \sinh y$.

[10] This follows immediately from the preceding. Since $dz = \cosh y\, dy$, and ⊙$A = 1$, integrating z from 1 to y at *NC* gives the area *A*⊙*NCA* = $\sinh y$ = the length of the arc z along *AC*.

[11] If *AR*:*BC* = ⊙*A*: ⊙θ, then ⊙θ is the fourth proportional. So (since ⊙$A = 1$), ⊙θ = *BC*/*AR*. Now ⊙*G* = ½ [⊙θ + ⊙*B*]. Thus since $AR = \sinh y$, ⊙$B = \cosh y$, and $BC = y$, we have ⊙G = ½ [$y/\sinh y + \cosh y$]. Leibniz would have obtained this expression as follows. The x-coordinate of the centre of gravity is given by $\int x\, dx / \int dz = \int \cosh^2 y\, dy / \int \cosh y\, dy$. Now $\int \cosh y\, dy = \sinh y$, and $\int \cosh^2 y\, dy = ½ [y + \cosh y \sinh y]$, and Leibniz's formula follows.

should assume any other configuration, the centre of gravity will descend less than if it is curved in our curve $CA(C)$.[12]

To find the centre of gravity of the figure enclosed by the catenary line and a straight line or lines: Take $\odot\beta$ to be half of $\odot G$, and complete the rectangle βAEQ; Q will be the centre of gravity of the quadrilineum $A\odot NCA$.[13] From the centre of gravity of any other space bounded by the catenary line and a straight line or lines is easily obtained. Moreover, from this follows the remarkable result that not only the quadrilinea such as $A\odot NCA$ are proportional to the arcs AC, as we have already noted, but that the distances of both centres of gravity from the horizontal through $\odot$, namely $\odot G$ and $\odot\beta$, are also proportional, since the former is always double the latter; and the distances from the axis $\odot B$, namely PG and $Q\beta$ are in a proportion that is simply that of equality.

To find the volumes and surfaces of bodies generated by rotation around any straight fixed axis of the figures enclosed by the catenary line and a straight line or lines: As is known, this is obtained from the two proceeding problems. So, if the catenary $CA(C)$ is rotated around the axis AB, the surface generated will be equal to a circle whose radius [squared] is equal in area to twice the rectangle EAR.[14] Other surfaces and also solids generated in this way can be measured no less easily.

I pass over many theorems and problems which, since they are either contained in the things we have said or can be derived from them without much effort, it seemed appropriate to treat briefly. So, for example, take two points of the catenary, such as C and ${}_1C$, whose tangents intersect at π, and from the points ${}_1C$, π, C drop the perpendiculars ${}_1C_1J$, πM, CJ onto the horizontal line AEE running through the vertex. Then the product of ${}_1JJ$ and AC minus the product of ${}_1CC$ and ${}_1JM$ will be equal to the product of ${}_1BB$ and $\odot A$.

Infinite series can also be usefully applied. So, for example, if the parameter $\odot A$ is unity, the arc AC or the straight line AR is denoted a, and the ordinate BC is called y, then it will be the case that $y = {}^1/_1a - {}^1/_6a^3 + {}^3/_{40}a^5 - {}^5/_{112}a^7$ etc., which series can be continued by means of a simple rule.[15] If, furthermore, the

[12] As noted by Hess and Babin (Leibniz 2011, 122), in modern terms the determination of the lowest possible centre of gravity is a variational problem. This would be solved today by an application of the Euler-Lagrange equation.

[13] Here, in modern terms the x-coordinate of the centre of gravity of the figure is ½ [$\int \cosh^2 y\, dy/\sinh y$], that is half the abscissa of the centre of gravity of the curve; and the y-coordinate of the centre of gravity of the figure is ½ [$\int y \cosh y\, dy/\sinh y$].

[14] As Parmentier observes (Leibniz 1995, 198, n. 35), here Leibniz is applying Guldin's theorem, according to which the surface generated is equal to the length AC multiplied by the distance traversed by the centre of gravity. Since this is at a distance AE from the axis of rotation, it traverses a distance of $2\pi AE$, so the surface generated is equal to $2\pi AE{\cdot}AR$.

[15] With $AR = a = \sinh y$, we have $y = \sinh^{-1} a = \int dt/\surd(1 + t^2)|_0{}^a$. Using the binomial theorem $1/\surd(1 + t^2)$ may be expanded as $1 - {}^1/_2t^2 + {}^3/_8\, t^4 - {}^5/_{16}t^6 + \ldots$ Integrating term-wise from 0 to a gives $y = {}^1/_1a - {}^1/_6a^3 + {}^3/_{40}a^5 - {}^5/_{112}a^7 + \ldots$, as desired.

factors determining the catenary line are given, the rest can be obtained from what has been said. So, for example, if the vertex *A*, another point *C*, and the length *AR* of the intercepted catenary *AC* are given, one can obtain the parameter *A*⊙ of the line, or the point ⊙: for, since *B* is also given, draw the connecting line *BR* and, from *R*, the straight line $R\mu$, so that the angle $BR\mu$ is equal to the angle *RBA*, and the line $R\mu$ (when produced) will meet the axis *BA* (when produced) at the point ⊙ that was sought. [280/281]

And I believe the preceding remarks contain the essentials, from which other things concerning this line can easily be derived if need be. For the sake of avoiding prolixity, I refrain from adding demonstrations, especially since they will emerge on their own for those who understand the calculations of our new analysis that have been explicated in these *Acta*.[16]

[16] Johann Bernoulli's solution (*Solutiones Problematis Funicularii*...) is given by him in (Bernoulli 1691). Immediately following the present article giving Leibniz's solution are the solutions of Huygens (*Christiani Hugenii, Dynaste in Zülichem, solutio ejusdem Problematis*), on pp. 281–82, and Jacob Bernoulli (*Specimen Alterum Calculi differentialis*...), (Bernoulli, 1691b, on pp. 282–90).

18

On Solutions to the Problem of the Catenary or Funicular, and to other problems proposed by the learned I. B. in the *Acta* of June 1691[1]

G. W. Leibniz, *Acta eruditorum*, September 1691[2]

[435/436]

Having read the three solutions to the problem proposed by Galileo and revived by Mr Bernoulli,[3] I am very delighted that they agree with one another, which is an indication of truth that will persuade those who do not examine such things closely. Even if there is no time to compare everything, nonetheless their agreement on the chief point of the matter is manifest. We have all discovered the law of tangents and the rectification[4] of the catenary curve; together with the measure of curvature I once explained—in the *Acta* of June 1686, p. 489 (after introducing a new kind of contact, which I decided to call 'osculation')[5]—by means of the radius of the circle osculating the curve, i.e. that one of all the tangent circles which most closely approaches the curve and so makes the same angle of contact with a straight line as the curve itself.

[1] From the Latin: *G. G. L. De Solutionibus Problematis Catenarii vel Funicularis in Actis Junii A. 1691, aliisque a Dn. I. B. propositis*, September 1691, 435–9. (EPS 212–16; GM V 255–58.) Translated by Jeffrey McDonough, Lea Schroeder, and Samuel Levey.

[2] This article is the comparison that Leibniz had promised (in the preceding article) of the three solutions to the catenary problem, his own, Huygens' and Johann Bernoulli's; together with some remarks on recent articles by Jacob Bernoulli, some corrections to his own article on the loxodrome, and some remarks on the origins of the calculus in response to Jacob Bernoulli's comparison of his calculus with Barrow's infinitesimal method. He concludes with a request for Jacob Bernoulli to give his own opinion on his reply to Papin's remarks in the immediately following article, [**19**] below.

[3] Jacob Bernoulli had posed the problem at the end of his *Analysis problematis antehac propositi*… (Bernoulli 1690a). The three solutions are given under the title *Solutiones problematis a J. B. propositi* on pp. 273–82 of the *Acta* of June 1691.

[4] Leibniz uses the expression *extensionem in rectam* (extension into a straight line) to mean the rectification of a curve, the finding of its arc length.

[5] *Meditatio nova de natura anguli contactus* (Leibniz 1686a). (GM VII 326–29; French translation by Parmentier in Leibniz 1995, 118–25).

Leibniz: Journal Articles on Natural Philosophy. Richard T. W. Arthur, Oxford University Press.
DOI: 10.1093/oso/9780192843531.003.0019

The famous Huygens (noticing that the centres of these circles always fall on those lines (first discovered by him) whose given points are described by evolution),[6] decided to give some thought to this and to investigate the radius of curvature or osculating circle of the catenary curve, i.e. the circle that generates its curve by evolution, which is also given by [Johann] Bernoulli's solution. In Huygens' solution, moreover, the distance between the centre of gravity and the axis is given as well; in Bernoulli's solution and mine, not only is the distance of the same thing from the axis given, but also its distance from the base or another straight line, and so is the determination of the centre of gravity, and likewise the quadrature of the catenary figure. In my solution, to these I added the centre of gravity of this figure ôr area as well. Mr Huygens shows the construction of the line from the supposition of the quadrature of the following curve, $xxyy = a^4 - aayy$;[7] Mr Johann Bernoulli and I reduced it to the quadrature of the hyperbola; he did this by very cleverly employing the rectification of the parabolic curve, while I did it by finally reducing the whole thing to logarithms, thereby obtaining *the most perfect means not only of expression but also of construction in transcendentals.* So, if one assumes or has a unique constant ratio even just once, concerning the rest an infinity of true points can be exhibited using common geometry without the further intervention of quadrature or rectification. One will perhaps enjoy seeing in my construction the marvellous and elegant concordance between the catenary line and logarithms.

Furthermore, Mr Huygens (giving us the hope of a considerable simplification using a table of sines) has observed that the matter also reduces to the sum of the secants of arcs increasing uniformly by minima.[8] The same thing had also been noted by me, and it had come to mind that the determination of the rhumb line or loxodrome used in navigation depends on the same considerations,[9] which I remembered also having defined from logarithms already many years [436/437] ago. For that reason, I searched my old papers

[6] According to Huygens, 'If a string or flexible line is understood to be stretched around a line curved in one direction and one end of the string remains fixed while the other end is pulled away so that the freed part of the line always remains taut, then it is clear that some other curve is described by this end of the string. This latter is called the curve "described by evolution"' (Huygens, *Horologium Oscillatorium*, Part III; 1986, 74). The evolute of a curve, in short, is the locus of all its centres of curvature.

[7] That is, the quadrature $Q = \int y\,dx = \int a^2/\sqrt{(a^2 + x^2)}\,dx$. If we make the substitution $x = a \sinh z$, we get $Q = a^2\, sinh^{-1}\, x/a$, in agreement with Leibniz's result for $a = \odot A = 1$.

[8] Thus we can solve Huygens' integral $Q = \int a^2/\sqrt{(a^2 + x^2)}\,dx$ by making the substitution $x = a \tan\theta$, so that $dx = a \sec^2 \theta\, d\theta$, and $\sqrt{(a^2 + x^2)} = a \sec \theta$, giving $Q = \int a^2 \sec \theta\, d\theta$, 'the sum of the secants of arcs increasing uniformly by minima'.

[9] The rhumb line or loxodrome is the curve that cuts all the meridians of a sphere at a constant angle less than a right angle (the rhumb angle). The resulting curve is a spherical helix. Its use in navigation is that it describes the path of a ship sailing with a constant bearing in relation to the true north.

and finally presented the matter publicly in last April's issue of this year's *Acta* (p. 181).[10] As it happens, however, the illustrious Professor of Basel Mr Jacob Bernoulli, the reviver of the problem of the catenary, also added to his brother's solution in last June's issue (p. 282) a consideration of loxodromic curves in which he reveals many extraordinary things.[11] He also gives the construction of the loxodrome from the supposed quadrature of a line whose abscissa and ordinate are z and x, and whose differential equation according to the form of my calculus is $dx = trrdz : z\sqrt{rr - zz}$.[12] Indeed, when he sees how I have reduced the matter to the quadrature of the hyperbola or to logarithms, I believe he will acknowledge that a certain finishing touch has now been put on this inquiry, and that all that remains is to adapt it more to practical use and popular comprehension.

Here, though, I should note that certain things in my mentioned construction of the rhumb line, published last April, should be corrected. For on p. 181, at line 12, '$_1l_2l$' should be replaced by '$_1l_3l$,' and at line 25, '$_1d_3l$' should be replaced by '$_2d_3l$', and p. 182, at line 20, 'ôr to $\frac{e}{1} + \frac{e^3}{3} + \frac{e^5}{5}$ + etc.' should be replaced by 'ôr to $\frac{e-(e)}{1} + \frac{e^3-(e^3)}{3} + \frac{e^5-(e^5)}{5}$ + etc.' At any rate, what needs to be replaced is evident from the preceding.

What Mr Bernoulli maintains in last January's issue, p. 16, concerning the equality of certain parts of curves that are dissimilar to one another, is very elegant.[13] As for the line of finite magnitude that makes infinitely many loops (June issue, p. 283), I do not think that it is unbounded, since it is equal to a finite line and could be traversed in a finite time by an equable motion.[14] I also have difficulty with what he said in last January's issue, p. 21, namely, that no (general) rectification of a closed geometrical curve is possible.[15]

[10] *Quadratura Arithmetica communis sectionum conicarum*... (Leibniz 1691).

[11] Jacob Bernoulli, *Specimen alterum calculi differentialis*... (June 1691b).

[12] In his article Leibniz had derived the differential equation for the loxodrome as $dx = ab\ dh/\cos h$, where a is a constant, b is the tangent of the rhumb angle, and h is the angle of latitude. In Jacob Bernoulli's formula, r is the radius of the sphere and t is the radius of the parallel at latitude h. If we set $z = r \cos h$, then Bernoulli's formula reduces to $dx = -tr\ dh/\cos h$, in conformity with Leibniz's. Bernoulli had treated the logarithmic spiral insofar as it relates to the loxodrome, with which it would coincide if the earth were flat (283).

[13] This alludes to Jacob Bernoulli's *Specimen calculi differentialis in dimensione Parabolæ helicoïdis*..., (Bernoulli 1691a), 13–23, a study of the helicoïdal parabola.

[14] In his *Specimen alterum calculi differentialis* of June 1691, Jacob Bernoulli had noted that both the logarithmic spiral and the loxodrome loop around the pole an infinite number of times, even though they are both of a finite length. This led him to declare in a corollary after determining this length that 'Since this spiral loops around the centre infinitely many times, it is clear that there can be a finite straight line equal to an unbounded curve' (1691b, 283). Leibniz's concern is that Bernoulli has failed to distinguish the unbounded from the infinite: an infinite curve may still be bounded, as in this case.

[15] Bernoulli had calculated that the highest point on the curve would be given if $2rry - 2ryy - rlt = 0$. Substituting $\sqrt{(lx)}$ for y, when $l = rr/c$, this reduces to $2\sqrt{(cx)} - 2x = t$, 'which equation cannot be resolved geometrically, because the ratio of x to t, the arc to the tangent, is unknown' (20–1).

I know another illustrious man sought to prove by a similar argument that no indefinite quadrature of a closed geometrical curve is possible;[16] yet, to Mr Huygens no less than to me, it seems that the matter is not settled.[17] And, unless I am mistaken, there exist counterexamples to which arguments of this kind can nevertheless be applied. I hope no offence will be taken at these criticisms, which I have made out of a love of truth rather than a zeal for contradicting, since they detract nothing from the other excellent things he has said. I am certainly the kind of person who, most willingly and with no end of pleasure, sings the praises of men who have clearly acquired or will acquire merit in the Republic of Letters, considering this to be the most deserving reward owed them for their labours, [437/438] and which may also serve as an incentive to both them and others in the future. I cannot deny having been wonderfully pleased by what the celebrated Bernoulli, together with his most ingenious younger brother, has built upon the foundations of the new calculus I have introduced; and that all the more so because, with the exception of that most acute Scotsman John Craig,[18] I had not yet met with anyone who had used it. But through their brilliant discoveries I hope that a matter which I judge to be of the highest utility, and which I see is also recognized by them, will have its usage propagated more widely in the literature. Nor is there any doubt that by this method Mathematical Analysis shall come closer to perfection, and that transcendentals thus far excluded shall be subjected to it. It has been excellently noted by Mr Bernoulli that at every point of inflection the ratio between t and y, or between dx and dy, there is a maximum or minimum of all possible values. And above all I have no doubt that they will discover things that would be difficult for me to attain: for there remain things on which I myself cannot yet work the matter out with the desired brevity. And just as I came upon these meditations chiefly on the occasion of reading the writings of Pascal and Huygens, and was only able to arrive step by step at results that could not easily be deduced from them and could hardly have been hoped for beforehand, so I believe my results, such as they are, will provide an

He comments: 'I note in passing that from this it can also be shown that the indefinite quadrature of the circle, and in general the rectification of any geometric curve closed on itself [*in se redeuntis*], is impossible' (21).

[16] Here Leibniz is referring to Newton's *Principia*, Lemma 28, Book I, Section VI: 'No oval figure exists whose area, cut off by straight lines at will, can in general be found by means of equations finite in number of their terms and dimensions' (Newton 1687, 71; Newton 1999, 511).

[17] See Huygens' letter to Leibniz of 5 May 1691 (A III 5, 86), where counterexamples are given.

[18] John Craig (1663–1731) was a Scottish mathematician who, as Leibniz says here, was the first to acknowledge his new calculus in print, in his *Methodus figurarum lineis rectis et curvis comprehensarum quadraturas determinandi* (Craig 1685). A former student of David Gregory at Edinburgh, Craig went to Cambridge in 1685 to study with Isaac Newton, and became one of the circle of 'Scottish Newtonians'. See Beeley (2010) for an account of Craig and his relations with Gregory and Newton.

occasion for others to discover further, even more abstruse things. And I am truly grateful to the illustrious Bernoulli for offering, and continuing to offer, problems related to the catenary: for example, when the catenary is of variable thickness, when the cord is stretchable, when an elastic band is substituted for the heavy cord, and finally, on the shape of a sail; I wish I had the time now to discuss these problems with him, but I have been so distracted by various kinds of work that it is only with difficulty that I was able to find time recently to elaborate and put into order my solution—discovered already a year ago—to the problem he proposed, and this was also the cause of the delay.

Furthermore, since he wanted to conjecture (p. 290) the circumstances by which I might have arrived at these meditations, and whose writings I might have had in front of me as my main aids in this, I am happy to candidly reveal this, too. I was a complete stranger to the inner recesses of geometry, when, in the year 1672, I made the acquaintance of Christiaan Huygens in Paris. I acknowledge that, after Galileo and Descartes, it is certainly to this man that these published articles, as well as I personally, owe the most in these matters. When I read his *Horologium Oscillatorium* and then added the *Letters of Dettonville* (that is, of Pascal) and the works of Grégoire de Saint-Vincent,[19] I suddenly saw the light—unexpectedly to both me and others who knew that I was a novice in these matters—and I soon showed this with examples. In this way a great many theorems revealed themselves to me, theorems which were only corollaries [438/439] of a new method, some of which I later recognized in the works of James Gregory, Isaac Barrow, and others.[20] But I noticed that their origins were not yet sufficiently clear and that there remained something more fundamental by which someday that loftier part of geometry could ultimately be reduced to analysis, of which it had previously been held to be incapable. I published some elements of this a few years ago, being mindful of public utility rather than personal glory, and I might perhaps have been further on my way to the latter had I suppressed the method. But to me it is more pleasing to see fruit growing from the seeds I have scattered in other

[19] Grégoire de Saint-Vincent (1584–1667) was a Flemish Jesuit and mathematician, known chiefly for his work on the quadrature of the hyperbola. Under Huygens' prompting, Leibniz studied his *Opus geometricum* (Saint-Vincent 1647) near the beginning of his stay in Paris.

[20] James Gregory (1638–1675) was a Scottish mathematician and astronomer, author of the well-regarded *Vera circuli et hyperbolae quadratura* (1668) and other works. Isaac Barrow (1630–1677) was a distinguished English mathematician and theologian, who taught Newton and preceded him in the Lucasian Chair of Mathematics at Cambridge. On vacating the Chair in favour of Newton, he asked his former student to edit his *Lectiones Geometricæ* (Barrow 1670). For the nature and extent of his probable influence on Newton's conception of fluxional mathematics and its application to nature, see Feingold (1993), Arthur (1995), and Guicciardini (2009, 169–81). See Mahoney (1990) and Probst (2015) for a refutation of the extravagant claims made by J. M. Child (e.g. in his introduction to Barrow 1916, vii) for Barrow's invention of the calculus.

people's gardens too. For there was neither time for me to cultivate these seeds fully myself, nor was there a lack of other areas in which I would uncover new approaches, which I have always judged more admirable, and I have considered methods more valuable than particular problems, even though the latter are more commonly applauded.

Finally, I shall add one thing, even if it is out of place here. I would like Mr Bernoulli to deign to give a second opinion on my response to Mr Papin concerning the estimation of forces,[21] especially at the end where I seem to have uncovered the source of popular errors. In the most recent July issue, p. 321, he quite rightly insists that no force is lost that is not spent somewhere;[22] but force is different from quantity of motion; and besides the fact that the firmer the obstacle, the less power is lost to it, it is most certain that impediments can be diminished in any given proportion; that the impediments due to rubbing ôr friction are not proportional to the speed (as I indicated in my *Schediasma de resistentia*);[23] and that even though there is a resistance of the medium, nothing precludes us from imagining oscillations in a place exhausted of air, or in a medium as rare as you please. Lastly, the mind must abstract from variable circumstances in order to explore the very nature of the thing itself.

[21] Leibniz is presumably referring to his second response to Papin, 'On the Laws of Nature and true estimation of motive forces', appearing in the *Acta* immediately after this piece ([**19**] below); and not to his first response, 'On the Cause of Gravity', [**13**] above.

[22] Bernoulli writes there: 'Huygens denies this consequence, saying that some motion is frequently lost, which is nowhere expended. But I am of the contrary opinion: if some goes missing, that would be perpetually spent somewhere else, but sometimes in pressing against a firm obstacle, sometimes in destroying an opposing motion, so that when our pendulum bobs are moved in the same direction, I will be able to infer correctly that the motion lost is necessarily consumed in the pressing against the axis' (Bernoulli 1691c, 321).

[23] 'A Sketch concerning the Resistance of a Medium', [**9**] above.

19

On the Laws of Nature and true estimation of motive forces, against the Cartesians: a Reply to the arguments proposed by Mr P. last January in these *Acta*, p. 6[1]

G. W. Leibniz, *Acta Eruditorum*, September 1691[2]

Since various obstacles prevented a prompter response from being given, I hope my most distinguished adversary will in his courtesy readily indulge me. In order to deliberate with suitable brevity, I am now setting aside the inquiry with him about the cause of gravity. The primary question about the estimation of forces is, what forces does nature always conserve the same? Most people estimate force as the product of mass into velocity, ôr quantity of motion, as a result of which the Cartesians want [439/440] the same quantity of motion to be conserved in nature. I, on the other hand, have shown in the *Acta eruditorum* of March 1686, p. 161,[3] that, since it is conceded by most people, principally by the Cartesians themselves, that it takes the same power to lift one pound to four feet as four pounds to one foot, that the force

[1] From the Latin: '*G. G. L. De Legibus Naturæ et Vera æstimatione virium motricium contra Cartesianos: Responsio ad rationes a Dn. P. mense Janurarii proximo in Actis hisce p. 6 propositas*', *Acta eruditorum*, September 1691, 439–47. (ESP 217–25; GM VI 204–15.) Translated by Richard T. W. Arthur.

[2] A week before mailing this reply to Papin to Otto Mencke in Leipzig, Leibniz had written a (no longer extant) letter for Papin's confidant Johann Sebastian Haes, perhaps in an effort to establish a private correspondence with Papin, which was in fact achieved later in the year. Leibniz argues that Papin has assumed what he is supposed to be establishing, namely that the force (here equated with the resistance to the falling body) is 'to be estimated by the product of velocity and quantity of body, that is to say, quantity of motion'. Against this he argues that the measure of force should be a measure of the power to achieve an effect which is *conserved under repeated substitution*: it is a quantity preserving that power that remains intact in the effect, that is, under substitution of effect for cause. When this power is used up in setting the body in motion, as Leibniz takes himself to have shown, this power is not proportional to speed, but to the square of the speed.

[3] See text [3] above.

Leibniz: Journal Articles on Natural Philosophy. Richard T. W. Arthur, Oxford University Press.
DOI: 10.1093/oso/9780192843531.003.0020

cannot be estimated by the quantity of motion; and that a body of four pounds having a velocity of one degree does not have as much force as a body of one pound having a velocity of four degrees, since if the former could raise one pound to 4 feet, the latter could raise the same pound to 16 feet. To this argument of mine certain people, in trying to respond, were so puzzled that they seem not to have perceived the matter well enough, yet they conceded the estimation of power as proportional to the mass and the height to which a mass or weight could be raised. But Mr P. rightly saw that when this was admitted, the estimation of power as proportional to mass and velocity could not stand; and persuaded of this, he denies it in the *Acta eruditorum* of April 1689, p. 183, and offers a certain argument of his own for velocity on a par with mass as proportional to force.[4] Responding in the *Acta eruditorum* of April 1690, p. 228,[5] and seeking to go into the matter in more depth, I taught that from the contrary opinion there follows an inequality between cause and effect, and even a perpetual motion, which seem absurd. I defined those things to have unequal forces for which, if one of them could be substituted in the place of the other there would arise a perpetual motion ôr an effect more powerful than the cause; and the substituted thing would have a *greater power*, that for which it is substituted, a *smaller* one. I showed, moreover, that by employing a certain mechanism, if a globe of four pounds having a velocity of, say, 1, is assumed to transfer all its force into a globe of 1 pound, and ought therefore according to the common opinion to receive a velocity of 4, there will arise an effect that is more powerful than the cause, ôr a perpetual mechanical motion. Here I call something *more powerful* if in its effect is contained another thing (an *inferior*), or else its effect and something else besides; that is to say, in this context more *powerful* and *inferior*, or *greater* and *smaller*, are when the (formal or virtual) contents of the latter are in the contents of the former, and something else in addition. Thus that which can lift a pound to 16 feet is more powerful than that which can lift it to 4; for it lifts it to 4 feet and a further 12 feet in addition.

The most distinguished P., in the January *Acta* of this year,[6] partly responds to my argument, and partly couples with it an argument of his own which requires a careful response in its own right. Noting both his perspicacity and candour, I will follow the foregoing that much more freely. Nevertheless, he seems only to respond to part of my argument. He candidly concedes that perpetual motion [440/441] would follow from the common opinion if the whole force of the globe of 4 pounds and velocity 1 could be transferred into

[4] See text [11] above. [5] See text [13] above. [6] See text [15] above.

the globe of one pound, but denies that this is possible. He says (*Act. erud.* of this year, p. 9) that I 'admit perpetual motion to be absurd, and the demonstration by the supposed transference is legitimate.' And then 'if he indicates to me some reason whereby the whole motive force, without a miracle, can be transferred from a larger body into a smaller one at rest, I will either grant perpetual motion, or hold up my hands in defeat'. Concerning the giving of this latter reason, I will say now that it is not needed at all for the force of my argument. For it was enough for me to show that 4 pounds with a velocity of 1 and 1 pound with a velocity of 4 could not have equal forces. For if *we suppose* that one can be substituted in the place of the other, perpetual motion will follow. I therefore have no need to show the way in which this substitution is effected *in actual fact*. But if someone denies this definition of mine of equal and unequal forces (which, however, Mr P. seemed to admit), then, so as not to dispute about terms, I ask only this, is it that nature does not in fact comply with it? or is it not that it takes care that this is never substituted for that *in actual fact*, when from the substitution of at least one of them for the other perpetual motion could arise? It is certain that experiments do not completely support it, nor is there any example to the contrary. With this conceded, I have no need that the whole force of the larger body be actually transferred into the smaller body; it suffices for me, for example (which Mr P. seems to concede) that the whole force of the smaller could be transferred into the larger. And so, if the whole force of 1 pound with a velocity of 4 could be transferred into a body of 4 pounds, and so according to the common opinion this would receive a velocity of 1, this falls into the absurdity (contrary to what was conceded) that one of them could be substituted for the other, of which the other could in turn be substituted for it, without a perpetual motion arising. And so it will follow that in the transferring of forces nature does not conserve the laws of equality with respect to effects. But if we suppose also that force is partly retained and partly transferred, we will fall into the same absurdity.

Perhaps there are some who either abandon every law of the equality of cause and effect (such as those who, admitting perpetual motion, think that as big an effect as you like can be produced from something however small), or at least deny that a perpetual motion ôr an effect more powerful than its cause is possible, but admit that an effect could be inferior to its cause. But I could scarcely believe that Mr P. would descend to this. For since he has conceded that it is not possible for the effect to be more powerful than its cause, admitting a cause [441/442] greater than its whole effect seems to be a way of avoiding the question rather than satisfying him, since both seem equally abhorrent to reason. It would also follow that the cause could not be restored again by being substituted for the effect, which, as is readily understood, would be very much contrary to the way nature works and the reasons for things. And with the effects always decreasing and never again increasing, the consequence would be that the very nature of things would continually decline with diminished perfection—almost like in morals where, according to the Poet, 'our parents' generation, worse than our grandparents', raised us good-for-nothings, etc.'[7]—and could never rise again and recover what was lost without a miracle. Which in physics is certainly abhorrent to the wisdom and constancy of our Founder. And it seems that one can accept as one of the first principles of this doctrine that (entire) cause and (entire) effect are always equipollent. And indeed I appeal to the good sense and candour of Mr P.: does it seems reasonable to him that instead of a power which could raise one pound to 16 feet there would immediately arise a power which could raise one pound only to 4 feet?—with the remaining power spent I do not know how, and so to speak annihilated, without any vestige or effect remaining. Which would definitely happen if instead of 1 pound with velocity 4 there could follow 4 pounds with velocity 1. But if this could indeed be brought to pass, then the whole effect when substituted could scarcely exceed a thousandth or hundred thousandth etc. part of what the cause could do. For if for 1 pound with velocity 1000 is substituted 1000 pounds with velocity 1 (which can happen according to the common opinion) the effect is reduced to a thousandth part, which seems completely absurd. And in general, if it is supposed that A is first given a velocity c and B a velocity e, whereas the bodies after transference and collision are A with velocity (c) and B with velocity (e), then according to the common rule defended especially by the Cartesians, that the quantity of motion must be conserved, that is, that $Ac + Be = A(c) + B(e)$, such numbers may be assumed that the same absurdities will arise.

But I would be surprised if Mr P. were given no cause for hesitation, not only by this, but also because he saw that the force of my demonstration was inevitable—unless he was denying that something was possible without any reason, that is, unless he was denying to the nature of things the faculty of bringing it about that the whole power of some larger body be at some time transferred, either immediately and directly, or mediately and circuitously, into some smaller body at rest. Certainly, the opinion that was rendered by him, that it could not stand if the latter were found possible, is in great trouble

[7] Here Leibniz is quoting Horace: 'aetas parentum, pejor avis, tulit nos nequiores mox daturos progeniem vitiosiorem [our parents' generation, worse than our grandparents', raised us good-for-nothings, soon to bear progeny even more degenerate]' (Horace, *Odes*, III, 6.)

and put in peril. To say nothing of the fact that it does not seem that a general law of nature should be suspended by such a condition, one also suspects it to be an evasion that he admits no postulates unless they are actually put into practice; as if someone [442/443] had refused to allow Archimedes' postulating some straight line to be equal to a curve because he could not exhibit one geometrically. And so I almost persuade myself that Mr P. will finally incline to our opinion in the end. If someone, however, could overcome the absurdities indicated earlier, I will hold in his grace what my most distinguished antagonist demands, 'to indicate a means by which nature could bring it about that the power of a larger body is transferred into a smaller one at rest'. But I could not adduce one. And even if it were conceded that the whole force of a smaller body could be transferred into a larger one, whether in motion or at rest, then *A* being moved, let us divide the larger body *B* at rest into parts smaller than *B*, each retaining the velocity of the whole *A*, and then successively transferring the power of any one of them into *B*, the whole power of the larger body *A* will be transferred into the smaller *B* at rest. To put it another way, let *A* and *B* be connected by a rigid line as long as suffices, and let us take a point *H* in it that is assumed to be set up in such a way that the composite does not even progress, but can rotate about the fixed point *H*. Let the latter be so close to *A* and so far from *B* that the speed achieved by *A* while circulating is as small as you like. Thus, *A* can be regarded as being at rest, or almost so, and almost the whole of its force, by the near dissolution of its striving ôr by the support of a rigid line, will be transferred to *B*. I employ rigid lines lacking mass, after the example of others who also imagine heavy bodies as points, and use other things of that kind as aids to demonstration, not at all spurning them, when it is a matter not of practice, but of investigating the reasons for things. And I have found that I have never deduced anything false from this.

When I was in Florence, I gave a friend yet another demonstration of the possibility of the transference of whole forces, etc., from a larger body into a smaller one at rest,[8] closely related to the very ones that the most distinguished P. very ingeniously thought up for my pleasure, for which I am indebted, and indeed give thanks befitting his sincerity. He would, however, soon try to reply, and on p. 11 argued concerning the example given by himself as follows: 'To this I reply that the validity of this argument cannot be denied if we suppose the lever perfectly hard and rigid, . . . but in fact there is no perfect hardness of this kind in the nature of things.' I hope, however, that when all is said and

[8] This friend was Rudolph Christian von Bodenhausen, on whom see the Introduction, as well as the second footnote to text [22] below.

done it will be noted that the force of the argument cannot be evaded so easily. For I hold for certain that the laws of nature and motion and must not be conceived in such a way that any absurdity should arise, even if bodies are supposed of the utmost rigidity, as I have also advised Dr Malebranche.[9] Nor, I believe, will any counterexample be adduced. [443/444] But why would it not also suffice for the effectiveness of the argument for perfectly rigid bodes not to be impossible, even if they were not given in actuality? Not to mention that according to the patrons of atoms such bodies are necessarily found in nature. But even if we concede that there are no perfectly rigid bodies, and indeed that there cannot be any, nevertheless, since elastic bodies that are able to rebound very quickly and fully are equivalent in effect to rigid ones, they appear the same as them down to as small a difference as you wish. So, if the lever itself were sufficiently rigid, or able to rebound fully and quickly enough, it could happen that the aberration from perfect rigidity would be smaller than a given one, and so all the force of our argument would remain, nor could a perpetual motion (at least in theory) be avoided unless the injunction to the opinion of the Cartesians were abandoned.

It remains for me to satisfy the opposing argument of my most distinguished antagonist. For with his fair-mindedness he seems to have been singularly impeded from not giving me a hand in demonstration, which he himself also hints at on p. 11 when he says: 'it is enough for me to show some deficiency in the argument, for since the demonstration proposed above proving the opinion of the Cartesians is most evident [it is like that, as we shall now see], then, as long as their adversaries' opinion does not enjoy the same certainty, it is clear enough that the former should be preferred'. Since, then, I have refuted the doubt raised against my demonstration, I will now proceed to refute the contrary argument. Which reduces to this:

> Those things which can overcome an equal resistance have equal forces;
>
> but a body *A* of 4 pounds with a velocity of one degree, and a body *B* of one pound with a velocity of 4 degrees, can overcome the same resistance;
>
> therefore the forces of bodies *A* and *B* are equal.

He proves the minor premise of this reasoning on p. 8 (reduced to formal logic) by the following pro-syllogism:

[9] For Leibniz's attempts to persuade Malebranche of the falsity of his collision rules, see the Introduction, and his first reply to Catelan (text [5]), and also the second reply by Papin (text [15]) above.

> Those things which can overcome an equal number of impressions of gravity can overcome an equal resistance;
> but bodies *A* and *B* can overcome an equal number of impressions of gravity;
> therefore, etc.

The minor premise of this pro-syllogism is then proved as follows: If the bodies *A* and *B* ascend in similarly inclined planes (or even if both are perpendicularly inclined), the times in which their force of ascent is consumed, ôr in which they reach as high as they can, will be in the ratio of the speeds, as is proved by Galileo's reasonings. But the speeds in our case are inversely as the bodies *A* and *B* (by hypothesis); therefore the times of ascending are also inversely as the bodies *A* and *B*; but the quantity of the impressions of gravity in conquering the ascent is in the compound ratio of the body on which the impression is made and the time during which the impression is made; [444/445] (since if the body and the time are alike divided into equal parts, the impression is equal in each part of the body and in each part of the time). And the ratio compounded of those things which are inversely proportional is the ratio of equality. So the quantity of the impressions, ôr the number of equal impressions of gravity on the bodies *A* and *B* is equal.

To this argument I reply by denying the minor premise of the principal syllogism, and I reply to the proof by denying the major premise of the pro-syllogism that goes as follows: Those things which can overcome an equal number of impressions of gravity can overcome an equal resistance. This proposition, I say, I deny, taking '*resistance*' to be quantity of contrary forces. And lest anyone think that I am denying it rashly or affectedly, it should be recognized that in it is contained the very point that is in question. For since the impression of gravity is nothing other than the degree of velocity impressed on each part, certainly if I allowed the resistance ôr contrary force to be measured by this impression, I would concede force to be estimated by the product of velocity and quantity of body, that is to say, quantity of motion. And I say these things indeed with regard to the argument advanced against me.

I, on the other hand—to reveal at last the reasons for my view—do not estimate quantity of resistance ôr of effect by the degrees of velocity, that is, by modal ôr incomplete entities, but by substantial ôr real ones. And in my judgement it is in the neglect of this that the principal mistake [πρῶτον ψεῦδος] on the part of my adversary consists. I judge those things to be EQUAL IN FORCE which can bring an equal number of springs equal in force to the same degree of tension, or which can raise the same number of weights to the same height above the prior situation of each of them; or even (if we prefer to translate the matter from concrete physics into pure mechanics), which can impress the same velocity on an equal number of equal bodies; or, finally, which can exhibit any thing endowed with power (as a *measure*) repeated an equal number of times. And I judge two things UNEQUAL IN FORCE to have a PROPORTION OF FORCES to another that is a proportion between the replications of *measure*, for example, between the numbers of springs equal to one another, or of weights to be equally tensed or lifted, or between the numbers of equal bodies receiving equal velocities to one another. By this way of reckoning the forces are reduced to a certain *measure* always congruent to itself and so much repeated, and it will turn out that the estimation made according [445/446] to one measure chosen arbitrarily, will also succeed according to any other measure; otherwise, nature would be without laws. But these things do not succeed, but instead are contrary to an estimation by means of replicated degrees of velocity, which I have shown do not agree with other unshakable ways of estimating that always agree with one another. And the true and intimate cause of this is that in this way, accurately speaking, no true and real measure is employed. For even if (as I want to be duly noted) I were to agree that three equal bodies having the same velocity have precisely three times as much power as one of them, since one and the same measure is here repeated three times—for a single body of the same kind and of the same quantity is repeated three times—I do not thereby concede that bodies having three degrees of velocity contain three times as much as a body equal to them having one degree of velocity, or that they thereby have three times its power. For even if it contains three times the degree of velocity, it still does not contain three times the quantity of body, but only one times it. From this it is clear that I do not exclude the role of velocity in estimating forces; for I show that whatever is finally employed for the determination of them, such as a spring of a given tension, a weight of a given magnitude and height, a body of a given mass and velocity, etc., one or many of them, if they can be presented and exhibited from the cause, they can also be exhibited from the effect, and vice versa. And whatever I finally assume as the *real measure* of forces, I always find agreement with the other cases.

But when a certain *modal* measure is assumed, for example, the replicating of the velocity without a replication of the body (namely, by holding that the forces of two equal bodies are as their velocities), we immediately fall into absurdities, and we either lose or gain part of the power without cause. Which could be of use in the example given, so that we do not depend too much on abstract things, or encroach upon the precepts of real metaphysics. From these things it is understood that the fact that up till now most people have not proceeded in this business correctly, arises from the lack of a *truly general Mathesis*, ôr universal *science of estimating*, which as far as I know has not been laid down, and of which here we give a specimen. But if for estimating the force we now employ the number of elastic pound weights to be overcome, or some other real effects agreeing with one another, the received opinion cannot stand, whereas mine will undoubtedly advance. At the same time all the absurdities adduced above will cease, and no cause will ever be substituted by another thing for which the other cannot in its turn be substituted, and no cause could ever produce what the entire effect could also produce in its turn; which do not hold in the adverse opinion. [446/447]

But it is not necessary to elaborate this at length, since it is easily seen by Mr P. and others who meditate on these things. I will be most grateful to learn whether some things remain on which the most distinguished gentleman considers himself not yet satisfied. And if he wishes to propose methodically and with swift feet, as is his custom, he will do me, as well as the cultivators of these letters, a great favour. In fact, I hope that in this way I have freed up what remains between us, and that a matter of such importance (by which the true laws of nature are to be established) is something we can pursue to its conclusion by continuing to confer between ourselves.[10]

[10] Leibniz followed up this suggestion by writing to Haes at the end of November 1691, setting the scene for the continuation of their exchanges through private correspondence. This news was well received by the journal's editor, Otto Mencke, who asked that, once the dispute had been brought to a conclusion satisfying to both parties, they should submit a synopsis of the controversy for publication in his journal. See the Introduction above. The subsequent discussion between Papin and Leibniz continued for years without either party managing to convince the other to his point of view. See O'Hara (forthcoming, Introduction §3) for an exhaustive (and exhausting!) account of the continuation of the controversy in their letters to one another.

20

General Rule for the Composition of Motions[1]

G. W. Leibniz, *Journal des Sçavans*, September 1693[2]

Let the straight lines *AB*, *AC*, *AD*, *AE* (etc.) represent different tendencies or particular motions of a body *A*, which must compose into a total motion; and let *G* be the centre of gravity of all the points of tendency, *B*, *C*, *D*, *E* (etc.). And finally let *AG* be prolonged from *G* to *M* in such a way that *AM* is to *AG* as the number of component or particular motions is to 1: then the composite motion will be *AM*.

That is to say, to speak more familiarly: if the moving body *A* coming from *A* had in one second of time come as far as *B*, in the case where it had been driven only by the motion *AB* (which I always suppose as uniform here); and again in the same way, if in one second it had reached as far as *C*, or *D*, or *E*, etc., in the case where it had been driven by one of these motions all by itself; now, supposing the moving body to be driven in the same time by all these motions together, not being able to go in the same time in several directions at once, it will go towards *G*, the centre of gravity of all the points of tendency, *B*, *C*, *D*, *E* (etc.), but all the more so as there are more tendencies: so that [417/418] in one second it will reach as far as *M*, if *AM* is to *AG* as the number of component or particular motions is to unity. Thus *the same thing will*

[1] From the French: '*Règle générale de la composition des mouvemens. Par M. de Leibniz*', *Journal des Sçavans*, Monday, 7 September 1693, 417–19. (ESP 303–05; Gallica 417–19; GM VI 231–233.) Translated by Richard T. W. Arthur.

[2] As part of his ongoing attempts to gain the support of French scholars for his physics, Leibniz had originally sent this piece on 6 May 1692 to his correspondent Paul Pellisson-Fontanier for publication in the *Journal des Sçavans*. Unfortunately, Pellisson died on 7 February 1693, by which time the piece had still not appeared in print. L'Hôpital's contribution to the journal of 15 June 1693 on the determination of tangents to focal curves prompted Leibniz to revise his manuscript in haste and resubmit it, together with the two applications of the rule in the article following. See O'Hara (forthcoming, Introduction, §3). In the article Leibniz outlines his method of combining various tendencies to move in the same body by computing the centre of gravity of the components. By splitting each tendency into components along two given perpendicular directions, he shows the vector addition of quantity of progress in each moment (i.e. momentum).

Leibniz: Journal Articles on Natural Philosophy. Richard T. W. Arthur, Oxford University Press.
DOI: 10.1093/oso/9780192843531.003.0021

happen to the moving body as would happen to its centre of gravity, if this body shares equally among the motions so as to satisfy perfectly all of them together. For with the moving body sharing equally among four motions, only a fourth part of the mobile can fall due to each of them, which would have to go four times as far in order to make as much progress as if the moving body all by itself had satisfied each tendency; but then the centre of gravity of all the parts will also go four times as far. Now, as there is no place for sharing, the whole will go like the centre of the shared motions, in order to satisfy each tendency in particular, as far as is possible without sharing. And it comes out just as if one had made the sharing, and reunited the parts in the centre, after having satisfied the particular motions.[3]

This explanation can take the place of a demonstration. But those who require it to be done in an ordinary fashion, will easily find it in pursuing what follows: If one draws through A two straight lines which are in the same plane as all the motions, and which make a right angle at A, one could resolve each of all these motions into two, taken on the sides of this right angle. Thus, the composition of all the motions on one of the sides will be the arithmetic mean motion multiplied by the number of of motions; that is to say, in order to have the distance between A and the point of tendency of this composite motion, taken on the side, it is necessary to multiply the distance from the centre of gravity of all the points of tendency on the same side by the number of tendencies. For one knows that the distance between A and the centre of gravity of the points taken on one and the same side is the arithmetic mean of the distances between A and these points, of whatever number they may be. I call the *arithmetic mean magnitude* among several magnitudes that which is made by their sum divided by their number, observing that that which is in a contrary direction is a negative quantity, whose addition is in effect a subtraction. Now, in order to determine the composite motion on each of the sides, taking the points of tendency on one side or the other of the right angle, it is necessary to multiply [418/419] their distances from the centre of gravity by the number of tendencies. From this it follows that the total motion, composed from the motions of these two sides, will determine that very thing. Thus the composition of several motions making an angle together in the same plane

[3] Thus if you take two tendencies to move of the same body and compose them by the parallelogram law, the centre of gravity of the 'points of tendency' will lie precisely half way along the diagonal, so that in this case the combined motion will be directly proportional to $AM = 2AG$ in a given time, along the direction of AG, where G is the centre of gravity. So AG represents the arithmetic mean of the two tendencies to move, and must be multiplied by 2 to get the total motion. Generalizing, the combined tendency for a body with n tendencies in different directions is $AM = nAG$, where G is the centre of gravity of all the points of tendency.

reduces to the composition of several motions in the same straight line, and of two motions making a right angle.[4] But if the given motions are not in the same plane, it is necessary to have recourse to three straight lines making [right] angles with each other.

It is well to remark that in this composition of motions the same quantity of progress[5] is always conserved, and not always the same quantity of motion. For example, if two tendencies are in the same straight line but in opposite directions, the moving body goes in the direction of the stronger one with the difference of the speeds, and not with their sum, as would happen if the tendencies carried it in the same direction. And if the two contrary tendencies were equal, there would be no motion. However, all this suffices, so to speak, *in the abstract*, when one supposes these tendencies already in the moving body; but *in the concrete*, when considering the causes which would have to produce them, one will find that in all there is conserved not only the same quantity of progress, but also the same quantity of absolute and entire force, which is different again from the quantity of motion. At another time we will give *two* very general and very important *consequences* which are derived from this rule.

[4] Effectively, what Leibniz has done here is to demonstrate the vector addition of directed motion where all the tendencies lie in the same plane. As he notes, this generalizes for non-planar tendencies, where their components in each of three mutually perpendicular directions must be added, so that these in turn compose to give the resultant overall tendency.

[5] 'Quantity of progress' is Leibniz's term for what Newton called 'momentum', the (vector) quantity of motion in a given direction; the conservation of directed quantity of motion had been demonstrated by Huygens, Wren, Wallis, and Mariotte in 1668–69. See Bertoloni Meli (2006, 227–40) for an illuminating discussion of the discovery of the correct impact rules.

21

Two Problems Constructed by Mr Leibniz, employing the general rule of the composition of motions that he just published[1]

G. W. Leibniz, *Journal des Sçavans*, September 1693[2]

Problem *A*. *To draw the tangent of a curved line which is described by tensed threads.*[3]

From the point *A* on the curve let there be described any circle whatever, cutting the threads at the points *B*, *C*, *D*, etc. Let the centre of gravity of these points, namely *G*, be found, and the straight line *AG* will be perpendicular to the curve; or, a straight line drawn through *A*, normal to *AG*, will be the tangent we are looking for. When the thread is doubled or tripled, one must consider there two or three points in only one place, somewhat as if these points take the place of others that are much heavier.

One can apply this construction not only to ordinary conics, to the ovals of Mr Descartes, to the co-evolutions of Mr von Tschirnhaus, but also to an infinity of other lines. Here is the reason for this, which served as the principle of invention. This is that one must consider that the stylus that stretches the threads could be conceived as having as many directions equal in speed to each other as there are threads; for it pulls them equally, and just as it pulls them, so

[1] From the French: *Deux Problemes construits par M. De Leibniz, en employant la regle generale de la composition des mouvemens, qu'il vient de publier*, *Journal des Sçavans*, Monday, 14 September 1693, 423–4. (ESP 306–07; Gallica 423–24; GM VI 233–234.) Translated by Richard T. W. Arthur.

[2] As announced in the previous article, these are applications of the rule Leibniz stated there for composing various tendencies to motion in a body. In the first he applies the rule to find the curve traced by a stylus whose motion is constrained by tensed threads, noting its relationship to previous results of Tschirnhaus, Fatio de Duillier, and the Marquis de l'Hôpital. In the second he notes how it can be applied in physics to combine solicitations such as the action of gravity and centrifugal tendency, as he had in his *Tentamen*.

[3] The idea here is that a curve is drawn by a stylus pushing against several tensed threads.

Leibniz: Journal Articles on Natural Philosophy. Richard T. W. Arthur, Oxford University Press. © Richard T. W. Arthur, Richard Francks, Samuel Levey, Jeffrey K. McDonough, Lea Aurelia Schroeder, and Tzuchien Tho 2023.
DOI: 10.1093/oso/9780192843531.003.0022

is it pulled by them. Thus, the composite direction (which must be along the perpendicular to the curve) passes through the centre of gravity of as many points as there are threads (by the new rule of the compositions of motion which the author just published in the preceding issue of the *Journal*). And these points, because of the equality of tendencies in our case, are equally distant from the stylus, and thus fall on the intersections of the circle with the threads. Mr von Tschirnhaus in his book titled *Medicine of the Mind*,[4] being the first to have researched this problem, gave Mr Leibniz[5] the occasion to get there; which he has done by taking a route which has this advantage, that the [423/424] mind can do everything without using calculation or diagrams.

Mr Fatio also got there on his own by a very beautiful route, and published it first.[6] Finally, on this subject the Marquis de l'Hôpital[7] has given the most general statement that one could wish for, founded on the new method *of the calculus of differences.*[8]

Problem Z. *When the same moving body is pushed by an infinite number of solicitations in the same time, to find its motion.*

I call *solicitations* the infinitely small endeavours, or *conatus*, by which the moving body is solicited or invited, so to speak, into motion; like, for example, the action of gravity, or centrifugal tendency, of which there must be an infinity in order to compose an ordinary motion. Find the centre of gravity of the *place* of all the points of tendency of these solicitations, and the composite direction will pass through this centre; but the speeds produced will be proportional to the magnitudes of the places. The places can be lines, surfaces, or even solids.

The problem which has just been resolved is of importance in physics; for nature never produces any action except by a veritable infinity of concurrent causes.

4 Tschirnhaus (1687). 5 Here Leibniz styles himself as 'M. de Leibniz'.

6 Nicolas Fatio de Duillier had corrected a mistake made by Tschirnhaus in his construction of caustics in the latter's *Medicina Mentis* in (Fatio de Duillier 1689), prior to his collaboration with Huygens. He later featured prominently in the history of the calculus by launching accusations of plagiarism against Leibniz in his (1699), a 24 page treatise to which Leibniz responded in (Leibniz 1700).

7 Guillaume François Antoine, Marquis de l'Hôpital (1661–1704) was one of the proponents of Leibniz's calculus in France. Using materials provided him by his tutor Johann Bernoulli, he wrote the first textbook on the Leibnizian calculus, (L'Hôpital 1696).

8 For L'Hôpital's solution see L'Hôpital (1693a and 1693b).

22

A Specimen of Dynamics for the disclosing of the admirable laws of nature concerning the forces of bodies and their mutual actions, and reducing them to their causes[1]

G. W. Leibniz, *Acta eruditorum*, April 1695[2]

Since I first made mention of founding a *New Science of Dynamics*, many distinguished people in various places have asked for a fuller explanation of this doctrine. As I have not yet found the time to compose a book, I will therefore present something here which may throw some light on the subject, something which will perhaps even be returned with interest if I can elicit the opinions of those who combine forceful thinking with elegance of style. I profess that their judgement would be most welcome, and hope that it would of benefit in advancing this work.

I have suggested elsewhere that in corporeal things there is something besides extension, indeed prior to extension, namely a force of nature implanted everywhere by the Creator. This force does not consist in a simple faculty, such as the Schools seem to be content with, but is further equipped with an endeavour ôr striving, one that will have its full effect unless it is

[1] From the Latin: '*Specimen Dynamicum, Pro Admirandis Naturæ legibus circa Corporum vires & mutuas actiones detegendis, & ad suas causas revocandis, autore G. G. L.*', *Acta eruditorum*, April 1695, 145–57. (ESP 357–369; GM VI 234–46.) Translated by Richard T. W. Arthur.

[2] As explained in the Introduction, Leibniz had composed a substantial treatise on his new science of dynamics while in Italy, the *Dynamica de potentia et legibus naturæ corporeæ*, and had left it in Florence in the hands of Bodenhausen for redaction and eventual publication, promising to send some corrections and additions. But he was never able to bring the project to completion, and, having announced his dynamics to friends in France and Germany (as well as in the *De primæ philosophiæ emendatione et notione substantiæ* of 1694), was under pressure to produce some specimen of it. He composed this *Specimen* in two parts, but the second part (GM VI 246–54), announced here as forthcoming in May, remained unpublished; as did two essays in French, both titled *Essay de dynamique*, written in 1692 and about 1700.

Leibniz: Journal Articles on Natural Philosophy. Richard T. W. Arthur, Oxford University Press.
DOI: 10.1093/oso/9780192843531.003.0023

impeded by a contrary endeavour. This striving is often apparent to the senses, but in my judgement is understood by reason to be everywhere in matter, even when it is not apparent to sense. But if we are not to ascribe it to God through some miracle, then this force must certainly have been produced by him in bodies themselves. Indeed, it must constitute the innermost nature of bodies, since it is the character of substances to act, and extension means nothing other than the continuation ôr diffusion of an already presupposed striving and counter-striving (that is, resisting) substance—so far is extension itself from being able to comprise substance.

It is beside the point that every corporeal action is through motion, and that motion itself comes only from motion, whether already existing in the body or impressed from elsewhere. For motion, just like [145/146] time, never exists, if you take things in a precise sense,[3] since a whole never exists when it does not have coexisting parts. And so there is nothing real in motion itself but that momentaneous thing which must consist in a force striving for change. Therefore, whatever there is in corporeal nature besides the object of geometry, that is to say, extension, reduces to this. And by this reasoning we can finally do justice at the same time both to the truth and to the teaching of the ancients. Just as our age has already rescued from scorn Democritus' corpuscles, Plato's ideas, and the Stoics' tranquillity about the optimum connection among things, so now the teachings of the Peripatetics about forms ôr entelechies (which have rightly been seen as enigmatic, and hardly perceived correctly by their authors themselves) are reduced to intelligible notions. Thus, I believe that a philosophy that has been accepted for so many centuries should not be abolished, but rather explained in such a way that it is consistent with itself (wherever this may be done), and moreover illustrated and augmented with new truths.

And it seems to me that this approach to our studies is best suited to a teacher's wisdom and of the greatest use for students. Otherwise, we will seem keener to destroy than to build, or will be tossed around by the winds of bold innovators, everyday uncertain in the face of their perpetual changes of doctrine. But when humankind has at last curbed the passion of the sects (which is stimulated by inordinate pride in innovating), and has established certain dogmas, it will take the first unhindered steps forward in philosophy no less than in mathematics. For if as a rule we leave aside the harsher things they say about each other, the writings of outstanding men, both ancient and

[3] Leibniz uses the Greek word 'ἀκρίβειαν', 'exactness': literally, 'if you reduce the matter to exactness'.

modern, usually contain a good deal that is true and good, which deserves to be extracted and arranged in the public treasuries of knowledge. If only people would rather do this than waste time on criticisms that only satisfy their own vanity! Certainly, although fortune has so favoured me with certain new things of my own that friends have often urged me to think only about these, I don't know why, but I nevertheless take some pleasure even in views alien to my own, judging each according to its worth, even though this varies. Perhaps the reason for this is that by doing many things we learn not to spurn any of them. But let us get back on track.

Active force (which you may well call *virtus*,[4] as some people do) is of two kinds. It is either *primitive*, which is in every corporeal substance as such (since I believe it is contrary to the nature of things for any body to be entirely at rest), or *derivative*, which is brought into play in various ways by primitive force as a limitation resulting from the conflict of bodies with each other. And primitive [146/147] force—which is nothing but the first entelechy[5]—corresponds to *the soul ôr substantial form*; but for that very reason it pertains only to general causes, which cannot suffice for explaining the phenomena. So, I agree with those who deny that forms should be employed in treating the particular individual causes of sensible things. This is worth pointing out since, while I am trying to give back forms their former right, as it were,[6] of revealing the sources of things, I would not wish it to seem as if at the same time I want to return to the verbal disputes of the Schools. Meanwhile a knowledge of forms is necessary for correct philosophizing, and only someone who has paid attention to such things can claim to have a satisfactory grasp of the nature of body; such a person will have understood that the crude notion of corporeal substance that depends only on the imagination is imperfect, not to say false—it is a notion that was carelessly introduced some years ago by an abuse of the corpuscular philosophy, which in itself is excellent and very true. This is also shown by the argument that such a notion of body does not exclude unconditional cessation or rest, and cannot offer reasons for the laws of nature governing derivative force.

[4] Here we are leaving *virtus* untranslated, whereas we use 'power' to translate *potentia*. Later in the essay Leibniz writes of *virtus* being used up in an effect like the raising a weight, the idea being that the live active force is used up and converted to dead active force, in which respect the latter is close to what we now conceive of as potential energy.

[5] Here Leibniz uses the Greek (perhaps to signal its Aristotelian provenance): *ἐντελέχεια ἡ πρώτη*.

[6] Leibniz also used this expression '*velut postliminio*', 'as if by postliminy', in his 1676 dialogue *Pacidus Philalethi* (A VI 3, 541/LLC 157). Postliminy is the right of returned prisoners, exiles, etc. to resume their former status in society.

Passive force is likewise of two kinds, either primitive or derivative. And, indeed, the *primitive force of being acted upon* or *of resisting* constitutes the very thing which, if rightly interpreted, is called *primary matter* in the Schools. This brings it about that one body is not penetrated by another, but makes an obstacle to it, and is at the same time endowed with a certain inertia,[7] so to speak, that is, a repugnance to being moved, and to this extent it does not allow itself to be impelled without some diminishment of the force of the body acting upon it. Whence the *derivative force of being acted upon* subsequently manifests itself in various ways in secondary matter.

But having distinguished and supposed these general and primitive considerations, by which we learn that a body always acts because of form, and is always acted upon and resists because of matter, it is up to us now to proceed further, and in the doctrine of *derivative powers and resistances*, treat the extent to which bodies act by various strivings, or in turn strive to act back on each other in various ways. For the laws of actions that apply to these things are not only understood by reason, but also corroborated by sense itself through the phenomena.

By derivative force, then, or the force by which bodies actually act or are acted upon by one another, I understand here only that which is consistent with motion (local motion, that is), and which in turn tends to produce further local motion. For I acknowledge that all other material phenomena can be explained through local motion. Motion is continuous [147/148] change of place, and thus requires time. But as a movable body that is in motion has motion in time, so it has a *velocity* at every single moment, which is greater the more space traversed, or the less time expended. Velocity taken with direction is called *endeavour*; whereas *impetus* is the product of the mass of the body and its velocity,[8] and its quantity is thus what the *Cartesians* are accustomed to call quantity of motion, namely, its momentaneous quantity—although, speaking more accurately, the quantity of motion itself, since it exists through time, arises from an aggregate of the products of the impetuses (whether equal or unequal) existing in the moving body during the time, multiplied by the times

[7] The Latin word Leibniz uses here is *ignavia*, meaning 'laziness, sluggishness'; but at this time inertia was only just receiving a more technical meaning, and in common usage has that same meaning of 'laziness, sluggishness'.

[8] The endeavour [*conatus*] is thus the directed velocity v, where v is represented by the increment in x, namely dx, in a moment of time dt, and thus proportional to dx and inversely proportional to dt (and v and dx are taken in the x direction, i.e. vectorially). *Impetus* is the product of this with mass m, namely mv. I have translated moles by 'mass' as opposed to 'bulk'. The Cartesians thought of this as a purely passive extended stuff, whereas Leibniz conceived mass rather as a derivative passive force that comprises a resistance to a change of motion as well as to penetration or deformation.

taken in order.[9] In disputing with the Cartesians, however, we have followed their manner of speaking. But just as we can also distinguish (and this is convenient for the technical use of language) an increase that is now occurring from one that has occurred or is yet to occur, as an increment or element of increase; or, just as we may distinguish the present state of descent from the descent made, which is increasing; so we may discriminate the present ôr instantaneous element of motion from the motion itself diffused through a stretch of time, and call it *momentaneous motion*; so what is commonly called quantity of motion would be called the *quantity of momentaneous motion.*[10] And although we may readily adopt terms after their meaning has been assigned, until we have done so we must be careful with them if we are not to be misled by their ambiguity.

Next, as motion through a stretch of time is estimated from an infinity of impetuses, so in turn impetus itself (even though it is a momentaneous thing) is estimated from an infinity of degrees successively impressed on the same moving body; it too contains a certain element from which it can arise only when this is infinitely replicated.[11] Imagine a tube *AC* rotating with a certain uniform speed in the plane of this page about a fixed centre *C*,[12] and that a ball *B* lying in the cavity of the tube is freed from any chain or impediment, so that it begins to move by centrifugal force. It is evident that the beginning of the

[9] Leibniz conceived impetus to be composed from an infinite aggregate of elementary strivings [*nisus*], which would give $mv = \int mdv$. The quantity of motion through time is the aggregate of products of the impetuses at different times with elements of time taken in order (as described here), which would give the impetus through time (by a second summation or integration) as $\int mv\,dt$. Leibniz (and others at the time) conceived of instantaneous (or momentaneous) velocity as an element of space traversed in a given element of time, always working with ratios of such quantities taken in the same element of time. As a result, an endeavour or element of instantaneous velocity, dv—which Leibniz calls a 'solicitation' for a body of a given mass, as we will see below—is then an element of an element of space (ddx) traversed in the same element of time (dt), and is thus not the same as an acceleration (ddx/dt^2 or dv/dt).

[10] The technical term Leibniz introduces for the momentaneous quantity of motion is *motio*, as opposed to the usual Latin *motus*, meaning motion over time. (In French, *motus* is translated as *mouvement*, so *motio* can be translated as *motion*.) Thus *motio* would be the quantity mdv in a given moment dt, represented by an infinitely small straight line tangent to the curve at each point of the curve, while the motion diffused through time (*motus*) would be the aggregate of these *motiones*, an infinite sided polygon representing the curve, whose total quantity of motion would be its initial value plus the integral of the successive increments mdv through time.

[11] Thus in his derivation of the motion of celestial bodies in elliptical orbits in the *Tentamen*, Leibniz represented the differences in the radii within an element of time by dr, and the difference of the differences by ddr, which gives the measure of the centrifugal force. So the endeavour away from the centre, and the solicitation of gravity balancing it, are infinitely small in comparison with the change in angle $d\varphi$ or the infinitesimal motion along the tangent, $rd\varphi$. Unlike Newtonian acceleration, this solicitation has the same dimensions as the impetus along the tangent, but not the same order of infinity. Likewise the impetus along the tangent itself, even though momentaneous, may be regarded as the infinite replication (or integral) of *nisus* of the next lower order of infinity.

[12] There are no diagrams in this article, so readers had to construct the accompanying figures in their imaginations.

endeavour to recede from the centre, namely that by which the ball *B* in the tube tends towards the end of the tube *A*, is infinitely small with respect to the impetus which it already then has from the rotation, that is, the impetus by which the ball tends from the place *D* towards (*D*) along with the tube itself while keeping the same distance from the centre. But when the centrifugal impression proceeding from the rotation is continued for some time, through that very process there must arise in the ball a certain complete centrifugal impetus (*D*)(*B*), comparable with the impetus of rotation *D*(*D*). From this it is obvious that there is a twofold *striving*: namely [148/149] an elementary ôr infinitely small one which I call a *solicitation*, and one formed from the continuation or repetition of elementary strivings, that is, the impetus itself. I do not mean, though, that these mathematical entities are really to be found as such in nature, but only that they are useful for making accurate calculations by mental abstraction.

Hence *force* is also of two kinds: one is elementary, which I call *dead force*, because there is no motion in it yet, but only a solicitation to motion, like that of the ball in the tube, or that of a stone in a sling while it is still held by a cord; the other is ordinary force when accompanied by actual motion, which I call *live force*.[13] And centrifugal force is of course an example of dead force, and so is the force of gravity, ôr centripetal force; also, the force by which a stretched elastic body begins to recoil. But in a collision produced by a heavy body that has been falling for some time, or from a bowstring that has been recoiling for some time, or from some similar cause, the force is live, and arises from infinity of continued impressions of dead force.[14] This is what Galileo meant when he said in an enigmatic turn of phrase that the force of impact is infinite, that is, if it is compared with the effort of gravity taken simply. Even though impetus is always accompanied by live force, however, these two differ, as will be shown below.

Live force in an aggregate of bodies can again be understood to be of two kinds, namely *total* and *partial*; and in turn, *partial force* is either respective or

[13] As explained in the Introduction, we have abandoned the long tradition of rendering *vis viva* as 'living force', and instead translate it as 'live force'. This is by analogy with a live cable, which is no more made a living being by its electric current than a mere aggregate is made a living being by simply having *vis viva*.

[14] As François Duchesneau (1994, 221–4) has explained, there are therefore two systems of integration involved. Dead force is a solicitation at the point where motion is about to begin, a virtual motion, and it consists in an infinite repetition of strivings mdv constituting the impetus $mv = \int mdv$, as explained above. But live force is the integral of this through time, $\int mv\, dt$. Thus it is proportional to distance; and if equal elements of velocity (i.e. *conatus*) are produced in equal times (as they are in the case of falling bodies), dt will be proportional to dv, so live force will be proportional to $\int mv\, dv$, and thus to mv^2.

directive, that is, specific to the individual parts or common to them all. *Respective* ôr *specific force* is that by which the bodies comprising the aggregate can act on each other; *directive* ôr *common force* is that by which the aggregate itself can consequently act on things outside it. I call it directive because in this kind of partial force the entire force of the overall direction of the aggregate is conserved. This is the only force that would be left if we imagine the aggregate to have suddenly congealed by the ceasing of the motion of its parts with respect to one another. Hence the *total absolute force* is composed of the respective and directive force taken together. But these things will be better understood from the rules to be treated below.

As far as we can tell, the ancients only had a science of dead force. This is what is commonly called mechanics, dealing with the lever, the pulley, the inclined plane (to which the wedge and screw belong), the equilibrium of liquids, and suchlike. This treats only the primary endeavour of bodies on one another, before they have acquired impetus through acting. And although the laws of dead force can be applied to live force in a certain way, we need to be very [149/150] careful that we are not deceived by this, like those who, recognizing that dead force is composed from the quantity made from the product of mass and velocity, have confused this with force in general. For this happens in this case for a special reason, as I pointed out on an earlier occasion: for example, when bodies of various weights are falling, then at the very start of their motion, when the descent itself ôr the quantities of space traversed in the descent are still infinitely small ôr elementary, they are proportional to the speeds ôr endeavours to descend. But when some progress has been made and live force has developed, the speeds acquired are no longer proportional to the spaces already traversed in the descent—although, how the force should be calculated I have shown before, and will explain further below—but only to their elements. Galileo began the treatment of live force (although under a different name, even a different concept), and was the first to have explained how the motion of falling bodies is produced by acceleration. Descartes correctly distinguished velocity from direction, and even saw that in the collision of bodies what follows is what produces the least change in the prior conditions.[15] But he did not calculate the least change correctly, changing only the direction or only the velocity, whereas the change should be

[15] Leibniz alludes to what Descartes wrote to Claude Clerselier on 17 February 1645, in defence of his Rules of Impact: 'when two bodies having incompatible modes collide, there must be some change in these modes in order to render them compatible, but this change will always be the least possible' (AT IV 185). As Bertoloni Meli comments, 'It is remarkable that Descartes would hold such a general principle to be the basis of the impact rules and refrain from publishing it' (2006, 155).

determined from both of them considered at the same time. But he could not see how this ought to be done; it escaped him because, being intent on modalities rather than real things, it did not seem to him that two such heterogeneous things could be compared and considered at the same time—to say nothing of his other errors in this field.

Honoré Fabri, Marcus Marci, Giovanni Alphonso Borelli, Ignatius Baptista Pardies, Claude Dechales,[16] and other very clever men have made valuable contributions to the doctrine of motion, but still they have not been able to avoid these capital errors. As far as I am aware, Huygens, who has enlightened our age with his brilliant discoveries, seems also to have been the first to have arrived at the pure and unadulterated truth in this matter, and to have freed this subject from fallacies by means of rules of motion he published some time ago. Almost the same rules were also obtained by Wren, Wallis and Mariotte, all of whom have excelled in these studies, though in different measure. They do not share the same opinion about causes, however, so that even people who are outstanding in these studies do not always come to the same conclusions. It is clear, therefore, that the true foundations of this science have not yet been disclosed. But not everybody accepts what seems certain to me, namely that rebound or reflection proceeds only from an elastic force, that is, from an internal [150/151] counter-striving[17] of motion. And no one before me has explained the very notion of force. This is something that till now has always troubled the Cartesians and others, who were unable to comprehend that the [quantity of] motion or the sum of impetuses (which they hold to be the quantity of force) could turn out to be different after a collision than before, perhaps because they believed that by this very fact the quantity of force would also be changed.

When I was still young, I conceived the nature of body, along with *Democritus* (and also with *Gassendi* and *Descartes*, his followers in this matter), as consisting only in inert matter. At that time, I brought out a little book called *A Physical Hypothesis*,[18] in which I expounded a theory of motion

[16] The relevant works of Fabri, Borelli, Pardies, and Dechales have already been listed above. Johannes Marcus Marci de Cronland (Jan Marek Marci) was a Jesuit doctor and scientist from Bohemia, rector of the University of Prague, and official physician to the Holy Roman Emperors. One of his main works was *De proportione motus seu Regula sphygmica: Ad celeritatem et tarditatem pulsuum ex illius motu ponderibus geometricis librato absque errore metiendam* (1639). For an enlightening discussion see Bertoloni Meli (2006, 150–3).

[17] This is the same word, *renisus*, Leibniz had used in the second and seventh paragraphs above. The idea is that it is an element of motion produced in a given body in reaction to an endeavour to compress or stretch it.

[18] This is Leibniz's *Hypothesis physica nova*, submitted to the Royal Society of London (A VI 2, N. 41).

abstracted from a system, as well as a concrete theory applied to a system, which seems to have pleased many distinguished people more than its mediocrity deserves. There I showed that, supposing such a notion of body, every body colliding with another gives its endeavour to that body, that is to say, gives it directly to whatever obstructs it. For since at the moment of collision it endeavours to continue, and thus to carry the other body away with it (because of the indifference to motion or rest that I then believed in), this endeavour would have to have its full effect on the body collided with unless it were impeded by a contrary endeavour—indeed, even if it were impeded by it, in which case these different endeavours would only need to be composed together. From this it was evident that no reason could be given why the colliding body would not achieve the effect towards which it was tending, ôr why the body being collided with would not accept every endeavour of the one colliding with it. And so the motion of the body being collided with would be composed of its own original endeavour and the new external one it has received. Moreover, from this I showed that if body is understood as consisting only in mathematical notions—magnitude, shape, place, and their changes—and an endeavour to change only at the moment of collision; if explanations make no use of metaphysical notions, namely, an active power in the form, and an inertia ôr resistance to motion in matter; and if therefore the outcome of a collision had to be determined only by a composition of endeavours, as we have explained; then it would necessarily follow that the endeavour of the whole colliding body, even the smallest one, would be impressed on even the largest body it collides with, so that the largest body at rest would be carried away by even the smallest body colliding with it, without any retarding of its motion, inasmuch as no opposition to motion is involved in such a notion of matter, but rather only an indifference to it. This being so, it would be no more difficult to impel a large body at rest than a small one, and so there would be an action without a reaction, and there would be no way of estimating power, since anything could be [151/152] overcome by anything else. Since this and many other things of the same kind are contrary to the order of things, and in conflict with the true principles of metaphysics, I then came to believe (and rightly so) that in constructing the system of the world, the all-wise author of things would have avoided these intrinsic consequences of the bare laws of motion derived by pure geometry.

But after I had investigated all this more thoroughly, I came to see what a systematic explanation of things consists in, and noticed that my former hypothesis about the nature of body was incomplete. And I noticed that other arguments as well as this one confirmed that besides magnitude and impenetrability we ought to suppose something else in body from which a consideration of forces might arise. When the metaphysical laws pertaining to this are added to the laws of extension, there arise what I call the systematic rules of motion. These are, namely, that all change happens by degrees; that every action comes with a reaction; and that no new force is produced without diminishing the previous force, so that any body which carries away another body will be slowed down by it; and that neither more nor less power is contained in the effect than was in the cause. Since this law[19] is not derived from the notion of mass, it must follow from something else which is in bodies, namely force itself, which always maintains its same quantity, even if it is exercised by different bodies. So from this I inferred that besides purely mathematical things subject to the imagination, we must admit certain metaphysical things perceptible only by the mind, and that a certain superior, and so to speak formal, principle must be added to material mass. For not all truths about corporeal things can be inferred from logical and mathematical axioms alone, that is, those concerning great and small, whole and part, shape and situation; but in order to provide reasons for the order of things, other axioms must be added concerning cause and effect, and action and passion. It does not matter whether we call this principle form, or entelechy, or force, provided we remember that it can only be intelligibly explained through the notion of forces.

I cannot agree, though, with certain distinguished men of today who, on recognizing this fact that the common notion of matter is insufficient, introduce God *ex machina*,[20] and remove all force of acting from things, as if in a kind of 'Mosaic philosophy', as Fludd once called it.[21] For even though I concede that they have clearly observed that there is no proper influx from one created substance into another, if the matter is determined in metaphysical rigour, and though I also freely admit that [152/153] every thing is always produced by God's continuous creation, nevertheless I believe that there is no natural truth in things whose reason is to be derived immediately from the divine action or will, but that there is always a reason implanted in things themselves from which all their predicates may be explicated. Certainly, it is agreed that God did not create only bodies, but also souls, to which there

[19] *Sic*; but Leibniz should surely have written 'since *these* laws *are* not derived...' etc., namely the systematic rules of motion he has just stated.

[20] Here Leibniz used the Greek for *ex machina*, ἀπὸ μηχανῆς.

[21] Robert Fludd (1574–1637) was a famous occultist of the early modern period. A prosperous London physician, Fludd was a follower of Paracelsus, but also a Cabbalist, Hermeticist, and Rosicrucian, whose writings provoked refutations by Marin Mersenne (1588–1648) and Pierre Gassendi (1592–1655).

correspond primitive entelechies. These things, however, will be demonstrated elsewhere by bringing out the specific reasons for them in greater depth.

Meanwhile, even though I accept that there is an active principle everywhere in bodies that is superior to material notions and so to speak, vital, still I do not agree here with Henry More and other men of singular piety and intelligence, who make use of some obscure *archeus* or hylarchic principle, even obtaining explanations of the phenomena in this way[22]—as if not everything in nature can be explained mechanically, and as if those who try this are seen as doing away with incorporeal beings, under suspicion of impiety; or as if it were necessary to follow Aristotle in affixing intelligences to the rotating spheres, or saying that the elements are driven up or down by their form: concise expressions, but useless for teaching us anything. I do not, as I say, agree with these things, and that kind of philosophy is no more to my liking than the theology of those who so firmly believed that Jupiter thunders or snows that they even charged those seeking more specific causes with the crime of atheism. In my judgement it is best to take a moderate path which satisfies piety as well as science: I acknowledge that all corporeal phenomena can indeed be obtained from mechanical efficient causes, but I understand that on the whole the laws of mechanics themselves are derived from higher reasons, and that in this way we use a higher efficient cause only for founding the general and less accessible principles. Once these have been established, then whenever we treat the more accessible and particular efficient causes of natural things we find no place for these souls or entelechies, any more than for superfluous faculties and inexplicable sympathies. For the first and most universal efficient cause should not itself enter into the treatments of special causes, except insofar as we are considering the ends which divine wisdom had in ordering things in this way, lest we miss an opportunity for singing the most beautiful hymns in his praise.

In fact, final causes can be employed fruitfully to great effect even in particular problems of physics (as I have shown in the particularly clear example of a principle of optics, warmly approved of by the celebrated Molyneux in his *Dioptrics*).[23] [153/154] This can be done not only so we can

[22] Henry More (1614–87) was a theologian and philosopher, usually regarded as one of the leading Cambridge Platonists. He was initially receptive to Descartes's philosophy, but later opposed several central Cartesian principles. His conception of space as infinite and absolute being sharing a number of attributes with God, and his insistence on the existence of an active principle in an intrinsically passive matter (his *Archeus*), were points on which he might have influenced Isaac Newton.

[23] William Molyneux (1656–1698) was an Irish philosopher of independent means who held various political positions during his career. He is known for his close association with John Locke, and for his formulation of what is now called 'Molyneux's Problem', discussed by Locke in the second edition of his *Essay on Humane Understanding* (Locke 1690). Molyneux's work on optics, *Dioptrica nova; A treatise of dioptricks in two parts*, was published in 1692.

better admire the very beautiful works of the supreme author, but also so we can at the same time determine this path, which is not equally apparent using efficient causes, or is so only on the basis of hypothesis. Philosophers have perhaps not yet sufficiently observed this use of final causes. And in general it should be held that everything in things can be explained in two ways: through the *kingdom of power* ôr through *efficient* causes, and through the *kingdom of wisdom* ôr through *final* causes; that God governs bodies as machines in the style of an architect, according to the *laws of magnitude* or *mathematical laws*, and indeed for the benefit of souls; whereas he governs souls, which are capable of wisdom, as his citizens, and as participating with him in the same society, in the style of a prince, or even a father, according to the *laws of goodness* or *morals*, redounding to his glory. These two realms interpenetrate each other everywhere, although the laws of each do not get confused with or interfere with one another, so that at the same time the maximum is obtained in the kingdom of power and the best in the kingdom of wisdom. But my task here was to establish the general rules for effective forces, which we can then use in explaining particular efficient causes.

Next, I arrived at the true way of estimating forces, and I have even reached exactly the same one by very different routes. One is a priori, from the simplest consideration of space, time, and action (as I shall explain elsewhere); the other is a posteriori, namely by estimating the force from the effect it produces in expending itself.[24] For here, by *effect* I mean not just any effect, but one by which the force is expended ôr used up, which you may therefore call *violent*. This is not the kind that a heavy body exercises in moving along a perfectly horizontal plane, since it always keeps the same force no matter how long such an effect is produced; although when this same *innocuous effect*, as I call it, is correctly employed, it is also in accord with our way of estimating; but we will set that aside for now.[25] The violent effect I have chosen is the one which is most capable of being homogeneous, ôr of being divided into parts that are equal and similar, like those of the ascent of a body endowed with gravity. For the raising of a heavy body to two or three feet is precisely two or three times the raising of the same body to one foot; and the raising of a body of twice the

[24] See in particular Leibniz's replies to Papin, texts [**13**] and [**19**].

[25] Leibniz's a priori argument proceeds from a consideration of unconstrained motion, which involves free as opposed to violent action. In the latter the force is used up in producing a particular effect; in free action, which is nothing other than the operation of force over time, force is conserved by a kind of continuous production of a virtual or formal effect. Leibniz gives an a priori argument to show that free action is conserved over any given time, and therefore is *vis viva*, since action is the product of this force and time. Leibniz tries to persuade De Volder with this argument (see Paul Lodge's introduction to Leibniz 2013), and also gives it in the later *Essay de Dynamique* of ca.1700.

weight to one foot is precisely twice the raising of a single weight to one foot; therefore the raising of a body of twice the weight to three feet is six times the raising of a single weight to one foot—supposing, that is, that heavy bodies are equally heavy whatever their distance from the horizontal (we may suppose this at least for the sake of exposition, for even though in fact things perhaps behave differently, here the [154/155] error is imperceptible). For it is not so easy to find homogeneity in an elastic body. Since therefore I wanted to compare different bodies having different speeds, for my part I easily saw that if there is a single body *A* and a body *B* that is twice *A*, while each of them has the same speed, then the force of the latter would be twice that of the former, since the latter must be supposed to have precisely twice what was once supposed in the former. For *B* has a body twice that of *A*, and the same speed, and nothing else. But if *A* and *C*[26] are equal and the speed of *A* is a single unit and that of *C* twice this, I saw that what is in *A* is not precisely doubled in *C*, since the speed is indeed doubled, but not, however, the body too. And I saw that a mistake had been made here by those who believed that force itself is doubled merely by this kind of doubling of a modality. This is just what I observed some time ago when I pointed out that (even after so many *Elements of Universal Mathematics* have been written), no one had yet stated the true *Art of Estimating*, which consists in the fact that one must finally arrive at something homogeneous, that is, at an accurate and complete reduplication not only of modes, but also of things.[27] No better and more illuminating example of this method can be given than that presented in the very argument above.

In order to obtain these results, then, I considered whether these two bodies *A* and *C*, equal in magnitude but different in speed, could produce certain effects that were equipollent with their causes and homogeneous with each other. For in this way things which could not easily be compared through themselves could at least be compared through their effects. But an effect should be assumed equal to its cause if it is produced by the expending or using up of the whole *virtus*,[28] where it does not matter how much time it takes for the effect to be produced. Let us then suppose that *A* and *C* are heavy bodies, and that their force is converted into an ascent, which will happen if at the moment when they have the two velocities mentioned above, *A* a single unit and *C* twice this, they are understood to be at the bottom ends of

[26] Here the text had '*A* and *B*', an obvious error.

[27] Here Leibniz is alluding to his argument about the estimation of force in his replies to Catelan (texts [**3**] and [**7**] above), and to Papin (texts [**13**] and [**19**]).

[28] As Leibniz noted above, *virtus* is another term for active force.

the vertical pendulums *PA* and *EC*. Now it is clear from what has been demonstrated by Galileo and others that if the body *A* with a speed of 1 ascends to a height $_2AH$ of one foot above the horizontal *HR*, then the body *C* with a speed of 2 can ascend (at most) to a height $_2CR$ of four feet. From this it already follows that the heavy body having a speed of 2 has four times the power of the one having one degree of speed, since it can bring about precisely four times as much by using up the whole of its active force: for in lifting one pound (that is, itself) [155/156] by four feet, it lifts one pound by one foot precisely four times. And by the same method it may be inferred generally that the forces of equal bodies are as the squares of their speeds, and therefore that the forces of bodies in general are in the compound ratio of their bodies directly, and of the squares of their speeds.

I have confirmed the same thing by reducing the contrary opinion—commonly accepted, especially by the Cartesians—to absurdity, namely, to its entailing perpetual motion. According to this opinion, forces are believed to be in the compound ratio of the bodies and the speeds. From time to time, I have also used this method to give an a posteriori definition of two *states unequal in active force* [*virtus*], and at the same time provide a sure criterion for distinguishing the greater from the smaller. For if substituting either one of them for the other results in a perpetual motion, ôr an effect more powerful than the cause, those states are not in the least equipollent, but that which was substituted for the other was more powerful, since it brought about something greater.[29] I hold for certain that nature never substitutes unequal forces for one another, but that the entire effect is always equal to the full cause; and conversely, that in our reasonings we can always safely substitute these forces by ones that are equal to them with complete licence, as if we had actually made that substitution in fact, and without fearing that this would result in a perpetual mechanical motion.

But if it were therefore true, as people commonly persuade themselves, that a heavy body *A* of 2 pounds (for such we will now assume) endowed with a speed of 1 unit, is equipollent with a heavy body *C* of 1 pound endowed with a speed of 2, we should be able to substitute one of them for the other with impunity. But this is not true. For let us suppose that *A* of 2 pounds has acquired a speed of 1 unit through the descent $_2A_1A$ from a height $_2AH$ of one foot; and now, while it as at $_1A$ on the horizontal plane, let us substitute for it (what they consider) an equipollent weight *C* of 1 pound and with a speed of 2. This will ascend to (*C*) at a height of 4 feet. Thus, simply by the falling of

[29] This was the method Leibniz used to try to persuade Papin, for example. See [**13**] and [**19**] above.

the two-pound weight *A* from a height $_2AH$ of one foot, and substituting equivalents, we have brought about an ascent of one pound to four feet, which is twice what it was before. Therefore we have gained that amount of force, that is, we have brought about perpetual mechanical motion, which is of course absurd. And whether we can actually bring about this substitution through the laws of motion is irrelevant, for a substitution among equivalents can safely be made even in the mind.[30] Even so, I have also worked out various ways in which it can actually be brought about, as nearly as you would like, that the total force of body *A* is transferred into body *C* which was previously at rest, but which is now (with *A* reduced to rest) [156/157] the only one in motion. From this it would happen that the two-pound weight with speed 1 would be exchanged for a one-pound weight with speed 2, if these were equipollent, and, as I have shown, this results in an absurdity. These are not idle considerations, nor are they mere verbal quibbles, but are extremely useful in comparing machines and their motions. For suppose someone had enough force—from water, animals, or some other source—to keep a heavy body of one hundred pounds in a constant motion by which it could complete a horizontal circle of diameter thirty feet in a quarter of a minute of time, and suppose someone else claimed that instead something twice its weight would complete only half the circle at a constant rate in the same time and at less expense, and suggested you would benefit from this, then you should know that you are being deceived and deprived of half your force.[31]

But now that we have disposed of these errors, we will set out the true and really admirable laws of nature a little more distinctly in the second part of this study, to appear in May.

[30] Again, Leibniz is here recounting his argument against Papin (see [**19**]), whom he had tried to persuade that in a thought experiment such as this one could demonstrate in principle how things must be independently of whether they could be realized in actual fact.

[31] This follows because the force (a measure of the ability to do work, for Leibniz) is proportional to the square of the velocity, and the heavier weight will be going half as fast, so that in the same time the ratio of $100 \times v^2$ to $200 \times (\frac{1}{2}v)^2$ is ½.

23

A Short Note on p. 537ff. of the December *Acta* of 1695[1]

G. W. Leibniz, *Acta eruditorum*, March 1696[2]

I would be unjust if I did not acknowledge that these more profound sciences owe a great deal to the meditations of the distinguished mathematician Professor Jacob Bernoulli of Basel, and I am obliged above all both to him and to his ingenious brother, Johann Bernoulli—now a famous professor at Groningen—for applying to various uses the kind of foundation of a certain higher analysis that I laid down, for extending it marvellously with their own discoveries, and for bringing it about that it is becoming more and more widely known and celebrated. I wish, however, to persuade that most celebrated man Jacob Bernoulli, whose most recent papers have incited me to this note, that it is very far from my intention to detract from his praiseworthy achievements.[3] I leave to him undiminished the glory of the discovery of the forms of the elastic sheet (on the basis, that is, of a very likely hypothesis).[4] I would not even have mentioned the theorems about the radii of osculating circles (even though they were not unknown to me) if I had not believed that their very simple origin, as well as that of other similar theorems, should be explained on the basis of a certain singular kind of differential calculus, when the opportunity arose. It did not enter my mind to use these theorems for the forms of the elastic sheet, because I had never directed my attention to finding

[1] From the Latin: *G. G. [L.]Notatiuncula ad Acta Decemb. 1695 pag. 537 Seqq.*, *Acta eruditorum*, March 1696, 145–7. (ESP 404–06; GM V 329–331.) Translated by Jeffrey McDonough, Lea Schroeder, and Samuel Levey.

[2] This note is a prompt reaction to Jacob Bernoulli's article in the December 1695 *Acta eruditorum*, *Explicationes, annotationes et additiones*, pp. 537–53. There Bernoulli had treated various problems: the form of an elastic sheet, rectifiable converging curves, the mean direction of tendencies, the drift of ships and the (nowadays so-called) Bernoulli differential equation. But he had expressed scepticism about Leibniz's claimed achievements with his infinitesimal calculus, and hinted that Leibniz had only dealt with the problems concerned after seeing Bernoulli's own solutions. He also proposed another challenge problem, providing a hint hidden in an anagram.

[3] On the first page of his article Bernoulli had expressed doubts about Leibniz's statements concerning the radius of the circle of osculation.

[4] This refers to Hooke's Law.

Leibniz: Journal Articles on Natural Philosophy. Richard T. W. Arthur, Oxford University Press.
DOI: 10.1093/oso/9780192843531.003.0024

those forms, not because the matter is not beautiful and worthy of investigation, but because with such an abundance of things to do, I did not wish to deal again with what I thought had been correctly dealt with by him, especially as I was also uncertain whether I could do it. And so he is wrong to seek the reason for this in the absence of theorems on osculation, especially since he himself admits that even after those publications neither Huygens nor I have so far given sufficient thought to these elastic lines. And even now, with the analysis of that outstanding man laid out, I cannot bring myself to enter into this admittedly very appealing field—for reasons that are more numerous than I would like.

Otherwise, I see that Mr Bernoulli hardly dissents from my opinions concerning constructions and I wish he would give more thought, if he finds the time, to the construction of transcendentals discovered algebraically by means of points. For that would be more analytical, even if we have not yet mastered this construction to the same extent as we have the reduction of quadratures to linear quantities. Regarding the number of roots in the osculation, I candidly confessed some time ago that I embraced Bernoulli's opinion after having investigated the matter more thoroughly. Since he is asking me for an example of a regular, rectifiable closed curve,[5] what occurs to me now is the epicycloid described by a fixed point on a circle that rolls on another circle. That this epicycloid is rectifiable has been shown by those most celebrated men Huygens and Tschirnhaus;[6] that it is a closed curve when the circumferences are commensurable, is shown by the construction itself. The Bernoulli brothers would do something wonderful if they would complete, either jointly or even in separate studies, the consideration they have begun concerning the shape of the sail.[7]

As for the mean directions, which I discussed in the *Journal des sçavans* of September 1693:[8] when the tendencies of a mobile point are infinite, I think the points can be assumed to have arbitrary tendencies in equal tiny intervals of time. Now, when the various points exert their tendencies, the progress of the common centre of gravity is also obtained from the progresses of the

[5] In the fourth paragraph of his article Bernoulli had expressed his belief that a closed geometric curve is not rectifiable.

[6] Huygens, *Horologium oscillatorium*, 1673, Pars III, Prop. X; E. W. von Tschirnhaus, *Curva geometrica, quae seipsam sui evolution describit*, (Tschirnhaus 1690b). As Hess and Babin explain (Leibniz 2011, 293), the length of an arc of a common epicycloid is $8(R + r)/m$, where R is the radius of the fixed circle, r that of the rolling circle, and $m = R/r$ is the ratio of the circumferences of the circles.

[7] Jacob Bernoulli had published articles on the shape of the sail: see his (1690a), (1693), (1694a) and (1694b).

[8] Leibniz is referring to his 'General Rule for the Composition of Motions', **[20]** above.

points, namely by comparing its situation before its progress with its situation immediately after the elementary progress of the points. But if we consider the tendencies of the impelled points, which are often different from the tendency of the impelling ones, the mean tendency received from them will be defined in the same way. And all these must be varied according to the situation in which they arise, but I expect extraordinary things in these matters to be published by this illustrious man swiftly and elegantly, and I think he should be asked publicly and by name not to withhold any longer his excellent studies on the motions of fluids and other things.

Regarding the controversies between Mr Huygens and Mr Renau,[9] the Engineer-general of the French navy, Huygens himself (I bear the loss of this truly great man all the more heavily as I was in particularly close contact with him and was well aware of his very great talents, among which intellectual strength and sincerity vied with each other) deigned to ask me for my opinion;[10] but at that time I did not yet have both sides of the dispute to hand. On the matter itself: some other time. It is rightly observed that the same wind impels a resting ship more strongly than a moving ship, and that sometimes this difference is not to be neglected. I also do not think that the declination [*dérive*] is the same when the force of the wind is different (as Mr Renau supposes), but rather that it is greater in proportion to the intensity of the wind.

The general method for constructing inverse tangents given on p. 373 of the August issue of the preceding year, I myself advertised merely as a mechanical one.[11] A very useful project is to reduce these inverse tangents to quadratures, or to separate the indeterminate quantities from one another. I can solve the *problem* Mr Bernoulli proposed concerning the differential equation $ady = ypdx + by^n qdx$,[12] and I reduce it to an equation whose form is $\ldots dv + \ldots vdz + \ldots dz = 0$, where the dotted parts are to be understood as

[9] This dispute began with Huygens' criticism of B. Renau d'Eliçagaray's anonymously published book, *De la théorie de la manœuvre des vaisseaux*, 1689. See A III 6, 103 and. A III 6, 319, and the explanations given there.

[10] Cf. Huygens' letter of 29 May 1694 (A III 6, N. 38). Leibniz had also discussed the matter with L'Hôpital, who made the missing writings available to him, as noted by Hess and Babin (Leibniz 2011, 294).

[11] In his *Constructio propria Problematis de Curva Isochrona Paracentrica* . . . , published in the Acta eruditorum of August 1694, pp. 364–75, Leibniz reveals 'a universal method for determining general solutions to differential problems, the neglect of which, if I am not mistaken, hindered the most distinguished gentleman from including all the curves of the kind he sought; so I will give a Mechanical method [i.e. a method of approximation] . . . by means of which any transcendental curves you wish to find, given differentially, can be drawn through a given point (when this can be done), and in as exact a manner as you could wish, even if not a geometrical one going through true points, like that announced above (for example in the curve of the sines), but only through points very close to the true ones'; p. 373.

[12] Bernoulli gives this equation at the end of his article.

arbitrary quantities given in terms of z. Such an equation, however, has been reduced by me in general to a quadrature, by a method that I communicated a while ago to friends and that I do not consider necessary to set out here. I am already satisfied if I have brought it about that the most astute Author of the problem is able to recognize a method that is (as I believe) not unlike his own. For I do not doubt that he too has arrived at the knowledge of such a method. I once attempted a great deal in this field, and achieved quite a few things, which now lie scattered among my papers, not readily accessible even to me—a fortune for a pauper—with the result that I seem to have them and not to have them at the same time. Nonetheless, my memory easily supplied the things above on the very day that I received the Leipzig *Acta* of last December, that is, yesterday, at the Brunswick Book Fair, where, between various distractions, I put these thoughts down on paper.

24

An Excerpt from a Letter of G. G. L. that he wrote to a friend in favour of his physical hypothesis about the motions of the planets, once inserted in these *Acta* (Febr. 1689)[1]

Acta eruditorum, October 1706[2]

A learned gentleman, who a few years ago in his astronomical work attacked my hypothesis,[3] did not satisfactorily grasp its strength and utility. For it had this most important advantage, that solid bodies move by a harmonic

[1] From the Latin: '*Excerptum ex Epistola G. G. L. quam pro sua Hypothesi physica motus planetarii olim* (Febr. 1689) *his Actis inserta, ad Amicum scripsit*', *Acta eruditorum*, October 1706, 446–51. (ESP 642–47; GM VI 276–280.) Translated by Richard T. W. Arthur.

[2] This piece has its origins in continuing controversies about the correct representation of orbits under a central force. Leibniz's theory in the *Tentamen* had been attacked by the Scottish mathematician David Gregory (see the following note). Meanwhile, Pierre Varignon, one of Leibniz's principal supporters in Paris, had published two articles (Varignon 1700) and (Varignon 1701) on motions under a central force. In subsequent correspondence with Leibniz he had pointed out that in the *Tentamen* Leibniz had incorrectly represented the centrifugal force (see the discussion in the Introduction above). Here Leibniz tries to answer both criticisms and corrects some misprints and other errors in the published *Tentamen*, as well as defending his assumption of the inverse square law of gravity by analogy with the intensity of spherical radiation emitted from a point, which will fall off in inverse proportion to the surface area of a sphere centred on that point. It represents a much-abbreviated version of the *Illustratio* that Leibniz had attempted to publish the previous year, and is not so much an excerpt from a letter to Varignon as a response promised him in their correspondence.

[3] Here Leibniz is referring to David Gregory (1661–1708), who in his *Astronomiæ physicæ & geometricæ elementa* of 1702 had criticized the *Tentamen* for its failure to account for the origin of gravity or for the orbits of the comets, and rejected the harmonic circulation as incompatible with Kepler's Third Law (Gregory 1702, 99–104). Gregory was an accomplished mathematician, familiar with Leibniz's *Nova methodus* since early 1686, and the nephew of the famous James Gregory (1638–1675). He had been ejected from his professorship at the University of Edinburgh on account of his Episcopalian beliefs, and, with support from Newton, had since been made Savilian Professor of Astronomy at Oxford University in 1691. Gregory had already begun drafting his commentary on the *Principia* as soon as he had received a copy in 1687, and especially after meeting with Newton in May 1694, enthusiastically espoused its doctrines against those of Leibniz. See Eagles (1977, 28) and Guicciardini (1999, 179–88) for discussion of Gregory's *Notae* on the *Principia*, and Beeley (2020) for an enlightening discussion of Newton's early Scottish circle.

Leibniz: Journal Articles on Natural Philosophy. Richard T. W. Arthur, Oxford University Press.
DOI: 10.1093/oso/9780192843531.003.0025

circulation in a fluid circulating in a precisely similar way, as if they were circulating only by their own impetus and their own gravity as though in an empty medium (that is, a non-resisting one); and on the other hand, so that the circulating fluid is not disturbed by the circulation of this solid, exactly as if the solid were not there, or were itself nothing but a part of the fluid; which cannot happen with any other circulation. Thus even if the solid and the fluid are not moved by conspiring circulations, nonetheless after a long stretch of time they are finally reduced to a conspiring harmonic circulation, by which they are not obstructed or disturbed by one another. From this it is also understood that the objections of this author against vortices ôr deferent fluid orbs do not hit home against this kind of circulation.

Furthermore it is easily grasped that the law of harmonic circulation followed by this fluid is conserved only in the same orb, so as to obtain agreement of the orb with its planet; and this is that much easier because the thickness of one orb—namely the one in which the planet turns if it is carried along with the whole planetary vortex—is negligible. But in the whole vortex, carrying different planets, where the median motion of the planet can be assumed to be as if in a circle instead of its whole elliptical motion through the orb, it must be said that to obtain the Keplerian law of periodic times, this also must arise from the very agreement of the free motion with the vortical motion, since this composition of the free impetus together with gravity also delivers the equilibrium itself of the vortical motion, as is explained more fully elsewhere.[4]

It should be observed, however, that even if nothing needs changing in the matter itself, still something in [446/447] our exposition needs changing for the better, by which the concordance of the truths will appear more absolutely. It should be said, namely, that the new paracentric impression of the planet that is harmonically circulating and at the same time gravitating toward the Sun or some other centre, consists of a conflict of gravitation and a single centrifugal endeavour; that is, not a double one, which occurred by my taking the term in an unsuitable sense, whose correction I think useful, so as to make the words agree as well as possible with the things concerned. At any rate,

[4] Here Leibniz is referring to Newton's composition of the orbital motion from the inertial motion along the tangent combined with the centripetal force of gravity, where the latter is in equilibrium with the centrifugal force. (See Bertoloni Meli (1993, 189–90) for an analysis, and defence of the thesis that Newton does indeed appeal to his Third Law to explain this equilibrium, thus committing himself to the reality of centrifugal force, contrary to the modern understanding.) Leibniz seems to be arguing that since the motion of a body harmonically circulating in a harmonically circulating fluid will be effectively inertial, then Kepler's Third Law should also be derivable from his composition of the motion from harmonic circulation and a contest between gravity and centrifugal force along the radius.

gravitation constitutes a new solicitation of motion towards the centre, whereas the centrifugal endeavour of the circulating body constitutes a new solicitation of receding from the centre, with both of them varying with the distance from the centre; and the total endeavour resulting from this consists in the difference between these endeavours, and follows the direction of the prevailing one. Moreover, *the centrifugal endeavour of the circulating body* can be taken in two senses: either as that which the moving body exerts if the nearest preceding motion is conceived as being along the tangent of the circle, or as that which the moving body exerts if the nearest preceding motion is conceived as being along the arc of the circle itself. For in the case where it decreases to the infinitely many times infinitely small, the angle of contact should be neglected. The former centrifugal endeavour really takes place at the beginning of the calculation, and so it is an initial endeavour, indeed, not an enduring one; the latter centrifugal endeavour really persists, and so takes place in the process of circulation.[5] The former one, then, which is an initial endeavour, we call *tangential*, the latter, which endures, *arcual*;[6] and supposing an equal circulation in each case, the arcual is double the tangential; since the latter would be represented by the versed sine, the former by its double. Put more simply, it is preferable to take the term centrifugal endeavour in the sense of *arcual*, for in this way the exposition of the circulation of a planet (that is, the one begun earlier) will be more elegant and rounded.

But in order to understand the matter, let *AC* be the radius of the body moving around the centre *C*, and let the elementary arc of the circle described about this centre be *EAG*, bisected at *A*, so that the chord *EG* cuts the radius at right angles and bisects it at *B*. Let the rectangle *ABGD* be completed, and with *BH* in *AC* taken equal to *AB*, let the parallelogram *AHGF* be completed. Now let us suppose the moving body *I* is moved uniformly with a velocity represented by the elementary space *IA*, equal to *AD*, and that, coming in a certain element of time from *I* to *A*, it reaches *A*, [447/448] with the angle *IAC* a right angle, and there by an attraction begins to adhere to the radius at the point *A*; and with its own impetus effecting a rotation of the radius around *C*, it then describes the elementary arc *AG* in an element of time equal to the previous one. The motion *AG* could be understood as composed of the prior impetus *IA* or *AD* together with the centripetal solicitation *AB* (for it does not matter

[5] Analogously, in his *Dynamica* Leibniz distinguishes an endeavour along a curved path from one along a straight line, which latter he terms a *nisus* (a striving). Yet he also wants to claim that the endeavour to move at an instant is along a straight line, and that there can be no acceleration in an arbitrarily small moment. These claims are incompatible for non-inertial motion.

[6] 'Arcual' appears to be a coinage of Leibniz's, meaning 'along the arc'.

whether *AG* is understood as referring to the arc or the chord), and so the moving body, when it describes *AG* instead of *AD*, is retained in a circle by the force which is as *AB* or *DG*; that is, the centrifugal endeavour (equal to the retaining force) is as *DG*, the versed sine of the arc *AG*; yet this is the initial, or tangential, centrifugal endeavour. But if the moving body at *A* is supposed to have already been turning in the circulation for a while, it would come not along the tangent *IA* but along the arc *EA*, as if it were coming from *E* to *A*, with the radius *CA* describing the arc *EA*—that is, with a uniform motion and in the same element of time in which it was previously said to traverse *IA*. With these things supposed, the impetus which the moving body has at *A* is as *EA*, which (no matter whether this is a straight line or a circular one) does not differ in length in comparison with *IA* or *AD*. The impetus of circulation therefore has the same magnitude as it did before, but a different direction, namely the chord *EA* which, when produced, falls on *F*. This is why when the circulation is continued in an element of time equal to the previous one, along AG, it can be conceived as composed of the previous impetus, e.g. *EA*, or *AF* (which is equal to and in the same direction as *EA*), together with the centripetal solicitation *AH*, or *FG*, which is twice that of DG, the versed sine of the arc *AG*. Therefore, twice the versed sine of the circulation will represent the arcual centrifugal endeavour, which is double the tangential or initial one, as was to be shown. And although it is true that arcual centrifugal endeavours are also as the versed signs of the arcs, since they are double in proportion to the simple ones; in fact, however, in comparison with the tangential ones represented by versed sines, they are expressed by their doubles.

I myself at one time (in the said essay of 1689) understood the term centrifugal endeavour not in the sense that it in fact constituted a new or elementary impression on the planet for receding from the centre arising from the circulation, and opposed to the gravitation towards the centre with a new impression on the agent; but in the sense of the former initial or tangential endeavour, which occurred first at the fore; [448/449] from which there arose an unsuitable expression, by which we hindered ourselves by having obscured the concordance of truths. For while in the figure of the said essay we have called *PN* the centrifugal endeavour, now it behooves us to say (as we have just also done) that the conflict is between the solicitation of gravity and the impression of the double endeavour. But if, as is proper, we understand by the term centrifugal endeavour that new impression arising from the circulation which resists the new impression of gravitation, then of course the centrifugal endeavour designates that which takes place during the circulation, that is the arcual one. In which sense it is most conveniently and simply called the new paracentric impression (by which the velocity of approaching the centre or of receding from it is changed); or, what is the same thing, the element of paracentric velocity is the difference between the solicitation of gravity ${}_3ML$ ôr $2aa\theta\theta$: rr (see §§15 and 19 of the said essay) and the centrifugal endeavour $aa\theta\theta$: r^3, namely the arcual endeavour which (by §§12 and 15) is expressed by double the inverse sine *PN*. Thus it was simpler in this case for the centrifugal endeavour to have been designated as arcual. But I suspect the occasion for taking one term instead of the other arose from the fact that I measured *the outward force* (and the centrifugal force is of this type) by the perpendicular from the point to the nearest preceding tangent; correctly indeed, but the tangent ought to have been understood as the line in which the direction was the nearest preceding one. Namely, *in the figure presented* (Table X, Fig. 3), a perpendicular must be drawn from the point *G* onto the continuation of *EA* (representing the direction of the preceding circulation), that is to say, onto *AF*; not onto *AD*, unless the circulation begins at *A*, in which case *AD* is the continuation of the preceding direction *IA*. Moreover, the difference between the perpendicular from *G* onto *AF* and *FG* is incomparable. I, on the other hand—even if not in the estimation of the matter itself, but in the term—generally considered the tangent *AD* at *A*, and the perpendicular to it, *GD*. Hence no error was produced, but only a lack of refinement, which I am now helping to remove. But the occasion for my noticing this was provided by the elegant meditations of the celebrated Varignon on centrifugal endeavour, even if I am looking into it in another way from him; since otherwise I had hardly been thinking at all about these things. [449/450]

Hence, in the said essay, in Paragraphs 11, 12, 15, 21, 27, and 30, instead of double the centrifugal endeavour one puts simply centrifugal endeavour; and instead of centrifugal endeavour taken simply, one puts its half. The same occasion will serve to emend certain typographical errors or mistakes of the pen, or even in the writing, from the same essay. Namely, in the *Acta* of 1689, p. 86, l. 4: 'sine' is to be omitted; p. 87, 8 lines counting from the bottom of the page: 'inferred from the repeated endeavours' is to be omitted; p. 90, in the second last line, for ${}_3M$ read *MR*, and in the last line, for ${}_3MH$, $H\odot$, and $\odot{}_3M$, read $\odot H$, H_3M, and ${}_3M\odot$; and p. 91, around the middle, for *BF*, *e*, read $\odot F$, *e*; and in the fifth line from the end, it will be useful to insert a comma between $-2aaqr\theta\theta$ and bbr^3. Also in the figure of the essay certain mistakes have been made. For in the line $\odot M_2$, D_2 should be put instead of D_1, and in the same figure, a point *G* should be put between M_2 and T_2 so that a straight line from *G* to M_3 is made parallel to the straight line from M_2 to *L*. And finally, at the bottom of the figure, to the three points Ω occurring in ascending order to the right, these letters, M, ${}_1G$, ${}_2G$, are to be ascribed in order; thus straight lines are produced, the later of each being greater than the earlier, namely these three: ΩM, $\Omega_1 G$, $\Omega_2 G$.

Already at the time when I had just published the essay so often cited, I had already communicated to some friends the reason [*causam*] why the gravitational attractions are in inverse square ratio to the distances (from the centre of the attracting heavy body). And I remember that when I explained this in Venice to that very celebrated gentleman, Michel Angelo Fardella, now a well-deserved professor in Padua with outstanding writings on philosophy and mathematics,[7] he replied that it had been shown in the same way by Geminiano Montanari,[8] a certain excellent mathematician, that the illuminations of objects are in inverse square relation to the distance from the thing radiating. Which is true, and has also been observed by others; and, since it does not concern the quantity ôr effect of radiation, there is equal reason for it when you do not conceive the rays as attracting (considering the thing abstractly and mathematically) or illuminating. Thus the elliptical motion of a planet is confirmed a priori by the law of gravitation shown by its causes. I, on the other hand, confirmed it in that essay from the phenomena, ôr a posteriori; that is, from the elliptical motion confirmed by observations I derived the law of gravitation and, [450/451] since it agrees so well with the mathematical arguments concerning radiations, I judged it to be that much more faithful to the hypothesis of the universe.

[7] Michel Angelo Fardella (1650–1718) was a Franciscan priest and professor at Padua. Leibniz had a lively intellectual exchange with him, which he recorded in *Communicata ex disputationibus cum Fardella* (A VI 4, N. 329). See Garber (2004) for a critical analysis.

[8] Geminiano Montanari (1633–87) was an Italian astronomer, lens-maker, and proponent of the experimental method in science. He succeeded Giovanni Cassini as astronomy teacher at the University of Bologna, and later taught in Padua. He drew an accurate map of the moon using an ocular micrometer of his own devising, and his observations on comets are cited by Newton in his *Principia*. This paragraph is an answer to Gregory's criticism that Leibniz offers no justification for the inverse square law.

25
Letters from Baron[1] von Leibniz to Mr Hartsoeker, with the replies of Mr Hartsoeker[2]

G. W. Leibniz and Nicolaas Hartsoeker, *Mémoires de Trévoux*, March 1712[3]

Editors' preamble

Mr Leibniz, asked by Mr Hartsoeker to give him [494/495] his opinion on *Les conjectures physiques* that the latter has published, initially sent Mr Hartsoeker some objections, to which this philosopher has responded in the *Eclaircissments*, without naming Mr Leibniz.[4] The dispute, far from concluding with the publication of the *Eclaircissements*, has become even more lively. Mr Leibniz found himself insensibly engaged in combatting the principles of the system of his adversary, that is to say, the perfect liquidity of one of his elements, and the indivisibility of the other. Maintaining against him that atoms are as impossible as a perfect liquid, he argued that the *cohesion* of

[1] Leibniz never himself used 'von Leibniz' or called himself a baron, and there is no evidence that he was ever appointed to any form of nobility.

[2] From the French: '*Lettres de Monsieur le Baron de Leibnits à M. Hartsoeker, avec les réponses de M. Hartsoeker*', *Mémoires de Trévoux* (more properly, *Mémoires pour l'Histoire des Sciences & des Beaux-Arts*), March 1712, 494–523; including: '*Lettre de Monsieur de Leibnits à Monsieur Hartsoeker*', 10 February 1711, 496–510; '*Lettre de Monsieur M. Hartsoeker to M. Leibnits*' [13 March 1711], 510–22; republished in the *Journal des Sçavans*, December 1712, 603–25. (GP III 516–21, and 522–7.) Translated by Tzuchien Tho.

[3] This article comprises Leibniz's letter to Nicolaas Hartsoeker of 10 February 1711, and Hartsoeker's reply to Leibniz of 13 March, 1711—that is, only one letter from each, contrary to what the title suggests. It was published in the *Mémoires de Trévoux*, with an introductory preamble and also a postface by the editors; and the whole article was subsequently republished in the *Journal des Sçavans* in December of the same year. Meanwhile an English translation had appeared in the *Memoirs of Literature* of London, 5 May 1712, 137 ff. The letters concern Leibniz's account of cohesion in terms of conspiring motions, Hartsoeker's notion of a vital first element, and his insistence on perfectly hard atoms, which Leibniz likens to Newton's hypothesis of action at a distance in regard to its unintelligibility in mechanical terms.

[4] The works referred to are Hartsoeker's *Conjectures Physiques* (Hartsoeker 1706) and his *Eclaircissements sur les Conjectures physiques* (Hartsoeker 1710).

Leibniz: Journal Articles on Natural Philosophy. Richard T. W. Arthur, Oxford University Press.
DOI: 10.1093/oso/9780192843531.003.0026

the parts of a body, which constitutes its hardness, had as its true cause the conformity of the motions that impel these parts: according to him, when these *conspiring motions* are disturbed by some accident, the parts lose their union and the bodies become liquid. Mr Hartsoeker did not at first understand what Mr Leibniz meant. The dispute occurred when Mr Leibniz sent the [495/496] first of the following letters to the Jesuit Fr. Des Bosses, residing in Cologne, to hand over to Mr Hartsoeker. Fr. Des Bosses, who is today regent of theology at Paderborn, has long been an intimate friend of the celebrated Mr Leibniz. His great learning and penetration, combined with all the virtues of a Christian friend, qualities that Mr Leibniz has recognized in the Jesuit, have made for this close friendship between them, despite the difference in their religion. Fr. Des Bosses has proposed that his celebrated friend make public his dispute with Mr Hartsoeker. We have obtained the latter's permission, and it is sure to provide great help to those who love to get to the bottom of the principles of physics.

Letter of Mr Leibniz to Mr Hartsoeker

You speak, Sir, as if you do not understand what [496/497] conspiring motion is, and ask if what I call by this name might not perhaps be the same thing as rest.[5] I respond, no, it is not. For rest does not tend to make or conserve the connection between parts that are at rest, and two bodies that remain next to each other do not for this reason make any effort to continue to remain together, whether they touch one another or not. But when there is a conspiring motion in their parts which is disturbed by a separation, some force is required to overcome that obstacle. Nor is it necessary in conspiring motions that the parts should not change their distances: they may very well change it, provided that this spontaneous change is something quite different from a violent change, which would make a separation and disturb these motions. The parts of the bodies resist separation, not because they have little tendency to separate, for in that case they would still resist if they were absolutely at rest, [497/498] contrary to what I maintain; but because they have a considerable motion that would have to be disturbed by the separation. If these parts tend towards separation by themselves, they aid anyone wanting to separate them;

[5] Harstoeker had written: 'A conspiring motion, you say, once it is established, is a state which tends to conserve itself and endure, as does all motion. But what you call conspiring motion, Sir, will it not perhaps be the same thing that others call rest?' (To Leibniz, 30 December 1710; GP III 511.)

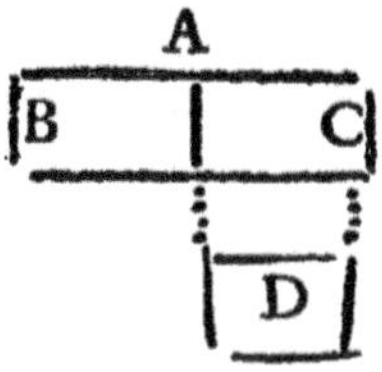

Fig. 25.1

but when they do not aid in this, it does not follow that they oppose one another, and some positive reason is needed for that.

I hold that force is required to drive a body out of its place, or to make it go faster than it would go by itself. But if [Fig 25.1] the body *D* tends to drive the body *C* from its place, the resistance of body *C*, which diminishes the speed of body *D*, contains nothing from which one might infer that body *B*, although nothing tends to drive it out too, should accompany body *C*; whether [498/499] the interval between *B* and *C* be great or small, or nothing at all. In order to produce this connection or accompaniment between *B* and *C*, then, there must be some other reason than rest, or the situation of the one with respect to the other; and as this must come from a mechanism, I have been unable to find it in anything but the conspiring motion common to some parts of the bodies *B* and *C*, which makes some parts pass from one body into the other by a kind of circulation, and which would have to be disturbed by the separation of the bodies.[6]

To say that conspiring motions are fictions[7] is to say in effect that every motion is a fiction. For how do you wish to make a motion, Sir, without there being some agreement among the motions of the parts? And the very nature of fluids that have been stirred up leads them to those motions that are the most mutually coordinated. You say, Sir, that your atoms have no parts, and you find it strange that I suppose one can conceive that an atom *A* has two [499/500] parts *B* and *C*. But are you not obliged to admit that an atom *D* may be conceived to go against the atom *A*, without going directly against the part *B*,

[6] Compare with the explanation of the cohesion of molecules according to modern band theory, where cohesion is due at least in part to the circulation of valence electrons shared by constituent atoms, even though (obviously) this is not a merely mechanical explanation.

[7] Hartsoeker had written: 'Atoms, you say, are fictions which one makes easily, but which are difficult to render intelligibly. But aren't conspiring motions, Sir, at least as great fictions, since it seems that there are no other motions than those by which bodies actually go from place to place with different speeds, and would it not therefore be better to take a solid and unshakeable foundation for once, and to maintain that there are atoms, that is to say, small solid masses, simple, homogenous, perfectly hard and without parts, and that these masses are so from all times by the eternal will of God, than to have recourse to incomprehensible and imaginary motions?' (To Leibniz, 30 December 1710; GP III 511–12.)

and this in such a way that it would carry *C* away with it but leave *B* in its place, if by chance *A* were not an atom or an otherwise solid body? There is then some ground for assigning parts in the supposed atom, and now one would need to assign causes for its 'atomicity', so to speak, that is to say, why *D* cannot carry C away with it without taking away *B* at the same time, and you would need to find a good glue to make one of these parts hold together with another, if you do not wish to have recourse, as I do, to conspiring motion.

If you invoke only the will of God for this, you are taking recourse to a miracle, and even a perpetual miracle: for the will of God works through a miracle whenever we cannot provide a reason for this will and for its effect through the nature of the objects. For example, if anyone should say that it is [500/501] God's will that a planet move circularly around its orbit, without anything aiding and conserving its motion, I say that this would be a perpetual miracle. For, by the nature of things, the circulating planet tends to depart from its orbit along the tangent, if nothing prevents it, and God would have to prevent this perpetually by a miracle if no natural cause does so. It is the same thing with the supposition of your atoms, for naturally the mass *C* will be carried away by the mass *D* without the mass *B* following if there is no reason that opposed this separation, and if you only seek this reason in the will of God, then you will only find it in a miracle.

It may be said in a very good sense that everything is a perpetual miracle, that is to say, worthy of admiration. But it seems to me that the example of the planet, which in circulating conserves itself in its orbit without any other aid than that of God, when compared with the planet retained in its orbit by the matter that always pushes it towards the sun, makes one well aware [501/502] of the difference there is between reasonable natural miracles, and miracles properly speaking, or supernatural ones, or rather (when these do not occur) the difference between a reasonable explanation and fictions where one has recourse to a miracle in order to support ill-founded opinions. This is thus what those people do who say, following the late Mr Roberval's *Aristarchus*,[8] that it is a law of nature that God has given in creating things, that bodies should attract one another. For without appealing to anything else to obtain such an effect, and without admitting anything that God might have done which could show how he attains this end, they resort to a miracle, that is to

[8] Gilles Personne de Roberval (1602–1675) was a professor of mathematics at the Collège Royale de France, and a founder member of the Académie des Sciences in Paris. Leibniz refers to his *Aristarchi Samii de Mundi Systemate* (1644), in which he supposed a mutual attraction between the Sun and the planets. For an account, see Auger (1957).

say, to something supernatural, and to a perpetually continued supernatural miracle, when instead it is a question of finding a natural cause.

You are right, Sir, to say that we should often acknowledge our ignorance, and that this is much better than launching ourselves into gibberish in order to give an account of things that we do not at all understand.[9] But it is one thing to admit that we [502/503] do not understand the reason for some effect, and quite another to assert that there is something for which we cannot give any reason, which is precisely to sin against the first principles of reasoning. This is just as if someone had denied the axiom that Archimedes employed in his book *On the Equilibrium*,[10] that a balance in which everything is equal on both sides remains in equilibrium, under the pretext that we do not sufficiently understand the thing, and that perhaps the balance would undergo change by itself without any reason.

Thus the ancients and the moderns who admit that heaviness is an *occult quality* are correct, if they understand by this that there is a certain mechanism that is unknown to them, by means of which bodies are pushed towards the centre of the Earth. But if their view is that the thing occurs without any mechanism, by a simple *primitive quality*, or by a law of God which produces this effect without using any intelligible means, this is an unreasonable occult quality, one that is so occult [503/504] that it is impossible that it could ever be made clear when even an angel, not to mention God, wished to explain it.[11]

The same goes for *hardness*. If someone admits that the mechanism that serves the foundation of hardness is unknown to him, he is right. But if he means that hardness comes from something other than mechanism, and if he has recourse to a primitive hardness, as do the defenders of atoms, he resorts to a quality that is so occult that it could not be rendered clear, that is to say, he resorts to something unreasonable and which sins against the first principles of reasoning by the admission it contains, that something natural happens for which there is no natural reason.

This is also the error of those who introduce an indifference of equilibrium, as if the will could ever be determined when everything is equal on both sides,

[9] Hartsoeker had written: 'And certainly, Sir, it seems to me that it is still much better to say and admit frankly that one cannot account for an extraordinary phenomenon, and thus to suspend one's judgement, than to make oneself ridiculous by launching into pure gibberish.' (To Leibniz, 30 December 1710; GP III 513.)

[10] Archimedes, *On the Equilibrium of Planes*, 1897.

[11] When Leibniz demanded redress from the Royal Society for John Keill's accusation that he had plagiarized the calculus from Newton, Keill brought these passages in the published letter to the attention of Newton, who was at that time President of the Royal Society and the running committee overseeing the complaint. This had the desired effect of provoking Newton to attack Leibniz rather than chastise Keill in the *Commercium epistolicum*. See e.g. Bertoloni Meli (1993, 186), Aiton (1972b, 145).

both internally and externally. Such a case never happens, and there is [504/505] always a greater inclination on one side than on the other, and the will is always inclined by some reason, or disposition, although it is never necessitated by these reasons. And I dare say that most of the mistakes made in reasoning come from not correctly observing this great principle, *that nothing happens without there being a sufficient reason for it*—a principle whose force and consequences have not been sufficiently considered even by Mr Descartes, and a number of other able people. This principle is itself sufficient to destroy the void, atoms, occult quantities, and even the first element of Mr Descartes, along with his globes and a number of other fictions.[12]

So, you can easily see, Sir, why God could not create atoms, that is to say, bodies hard in themselves, bodies of a naturally primitive hardness, bodies of an insuperable hardness, and for which there was no reason; just as he could not create planets that circulate by themselves in their orbits without there having been any [505/506] reason to prevent them from departing along the tangent. For some miracle at least must keep the planet in, or prevent the parts of a hard body from separating, if no mechanical or intelligible reason does so. Supposing we grant atoms but are far from admitting the void, this would not force us to resort to a first element, that is to say, a perfectly fluid matter. For why couldn't space be filled up by a matter that had different degrees of fluidity and tenacity, as I believe is the nature of all matter?

I also do not see why it is necessary for hard bodies to receive all their motion from fluid bodies and, above all, from a perfectly fluid mass, or your first element.[13] For all matter being equally susceptible to motion and equally incapable of deriving it from itself, nothing prevents the cause of its motion from giving it to the firmest bodies [506/507] as well as the most fluid. One could even say that the motion given to a few firm bodies could serve to account for the motion of many fluid bodies, and that consequently it is prior

[12] In his *Principia Philosophiæ*, Descartes had proposed that matter is particulate, and that it consists of three basic elements: that whose parts compose the large macroscopic bodies of experience (including the Earth, planets, and comets); the much smaller spherical 'globules of the second element', and the tiny, fast moving 'shavings' of indefinite size produced by the collisions of bodies of the first two elements (Part III, 48–52; AT VIII, 103–5).

[13] Hartsoeker had written: 'My first element is an infinite substance in which the atoms move from place to place, and from which they receive all their motion and direction; finally, it is the soul of the Universe, so that one may say that the Universe is like a great animal full of life and intelligence. "If the first element," you say, "is endowed with intelligence, and if it is a creature, it is necessary that this extended, impenetrable thing, capable of pushing atoms, be organic, in order to feel what it does, and to operate according to what it feels." I believe, Sir, that it is impossible for the human mind to decide how a substance should be made or constituted to act, to have intelligence, and to have an idea of what it does. It seems that the soul needs animal spirits and organs to think, and perhaps the first element needs atoms to act and exercise its intelligence.' (To Leibniz, 30 December 1710; GP III 514.)

in order. For a firm body put into a full fluid sets the whole of it into motion and produces a sort of circulation necessary to refill the place, without which there would remain a void behind the firm body. And this circulation forms a sort of vortex that has some affinity to the one that is conceived to surround the magnet. It should not be said that the universe is like an animal full of life and intelligence: for from this we might be led to believe that God is the soul of this animal; whereas God is a *supramundane intelligence* who is the cause of the world. And if the universe were boundless, it would be a mass of animals and other beings; but it could not be an animal.

Also, your first element is no more capable of life and intelligence [507/508] than any other mass, and since this body is not an organic one, it is not appropriate that it would have perception, which must always correspond with the actions of organs if you hold that nature should act with order and connection.

You say, Sir, that it is impossible for the human mind to penetrate how it is that a substance should have life and perception, and you are correct when it is a question of the details and the beginning of things. But you will perhaps also admit that this is more intelligibly explained in my system of Preestablished Harmony by conceiving that our substances naturally represent what happens in that portion of matter to which they are united.

I have sufficiently satisfied those who have objected to me that according to this there would no longer be *freedom*; for God, knowing what minds will freely choose in time, has accommodated bodies to this in advance. Mr Jaquelot, who offered me such an objection [508/509] in person, was satisfied with my response, as he admitted in his book against Mr Bayle.[14] He even clarified this with an elegant comparison. I have also responded in the same way to Father Lamy[15] and my response is in the *Journal des sçavans*.[16] Mr Bernoulli, when he was Professor at Groningen, has maintained theses

[14] Isaac Jaquelot (1647–1708) was a noted Hugenot proponent of rational theology, whom Leibniz had met in Berlin in 1702. The book referred to is Jaquelot's *Conformité de la foi avec la raison: ou défense de la religion, contre les principales difficultez répandues dans le Dictionaire historique et critique de M. Bayle*, Amsterdam, 1705. Jaquelot could not accept Leibniz's pre-established harmony as a solution to the mind-body problem, however, and criticized it in his later *Réponse aux Entretiens composez par M. Bayle, contre la conformité de la foi avec la raison, et l'Examen de sa théologie*, Amsterdam, 1707. See Woolhouse and Francks (1997).

[15] Father François Lamy (1636–1711) was a French monk of the Benedictine order. He had criticized Leibniz's notion of pre-established harmony in the second volume of his *Connoissance de soy-même*, Paris 1697. See R. S. Woolhouse (2001) and the introduction to Woolhouse and Francks (1997) for discussion.

[16] Leibniz published his response as 'Réponse de Mr Leibnitz aux Objections que l'Auteur du Livre de la Connaissance de soi-même a faites contre le système de l'Harmonie Préétablie.' *Supplément du Journal des sçavans*, 1709, 275–81. Reprinted at GP IV 577–90.

in which he had strongly defended my opinion about the pre-established harmony.[17]

Concerning the rest, the imperfections that are in the universe are like the dissonances in an excellent piece of music, which contribute to making it more perfect, in the judgement of those who sense the connection well. Thus we cannot say that God, in creating the world, has made an imperfect machine which develops in a poor way. It is true that there are machines in the world which do not always and from the beginning have every perfection of which they are capable.

I offer my thanks to you, Sir, for your good wishes for the [509/510] New Year, and I hope for your long and continued contributions to the growth of science, being with passion,

Sir,

Your most humble and most obedient servant,
Leibniz

Hanover, this 10th of February

Letter of Mr Hartsoeker to Mr Leibniz

Sir,

I do not know, Sir, if I have too limited a mind, or even if I have been too preoccupied with my atoms, to understand the arguments by which you try to prove and establish your conspiring motions. Matter is eternal, according to some pagans, or created by God, according to the moderns. If the former were true, nothing would prevent it from being divided into bodies [510/511] of perfect hardness, and to be so in itself and by its nature. But if matter was created by God, I ask you, Sir, if it might not have been created as he would have wished it to be, whether for an instant, or for some limited space of time, or for eternity, without employing only his will. If some mechanism is needed for this, I admit to you frankly, Sir, that I am unaware of it. For as for your conspiring motions, I still do not understand anything of them. A body may be at rest or in motion, and as the quantity of its motion is measured by the

[17] See Johann Bernoulli's letter to Leibniz of 11 February 1699, where he discusses the three main ways of explaining harmony, 'the first that of influence, the second that of [divine] assistance, and the third that of pre-ordination'; he rejects the first two, and embraces the third as 'the most elegant and most worthy of God' (GM III 568–69).

product of its magnitude and its speed, it has very little motion if it is very small and if it has very little speed; but as a body that has very little motion is very easily diverted, and that it can easily receive any motion it is given, how does it then come about, Sir, that the parts of a diamond, which have without doubt very little motion, if they have any at all, have such a connection [511/512] together that they make a body of the hardness we see? For me, I say that it has this hardness because it is composed of bodies of a perfect and insuperable hardness, just like all other things in this visible world, without excepting water, air, aether, and what could be even more fluid. Water is only fluid because the little bodies of perfect hardness that compose it are just hollow balls that the heaviness of the atmosphere cannot join together; except that when they touch each other too closely at their openings, they can then create the effect of little planes, and thus form what we call ice. And diamond is hard and endures for many centuries in the same state without any change only because the little bodies of perfect hardness, the little solid masses from which it is composed, are very strongly connected together by the atmosphere of the earth that weighs down on them.

If you do not, then, agree with me that these little extended masses, [512/513] solid and of an insuperable hardness, are the principle of all sensible bodies, I defy you, Sir, to explain in an intelligible manner the constant hardness of some, the fluidity of others, and so forth. Give me the materials if you want me to make you a building. For without that I could be the best architect in the world and yet not be able to construct any edifice. 'To say that conspiring motions are fictions,' you say, Sir, 'is to say in effect that every motion is a fiction.' But I reject this consequence. I know well, Sir that there is an infinity of bodies which have some agreement among their motions, but I say that there is no motion which could cause the hardness of bodies by itself; and certainly, Sir, when you say in your letter that 'the parts of the bodies resist separation, not because they have little tendency to separate, for in that case they would still resist if they were absolutely at rest, contrary to what I maintain; but because they have a considerable motion that would have to be disturbed by the [513/514] separation,' I must admit to you, Sir, that my mind is too small to understand any of this, let alone the following: 'if these parts tend towards separation by themselves, they aid someone wishing to separate them; but when they do not aid in this, it does not follow that they oppose one another, and there must be some positive reason for this.' Where is the considerable motion that the parts of a diamond could have which subsists for many centuries without any change? If you do not call motion something completely different from what the world understands by this name, then what do you call the tendency of the parts of a body to separate, or to unite and to be connected together? Finally, what is it that you want to say, Sir, by these words, 'if the parts tend towards separation by themselves'? It seems to me, Sir, to tell you the truth, that you use the words '*tendency*' and '*to be tending*', without attaching any idea to them. You say, Sir, 'if you invoke only the will of God for this, you are taking recourse to a miracle, [514/515] and even a perpetual miracle.' Let it be so, Sir, that I have recourse to this, just as you are obliged to have recourse to the continual existence of your conspiring motions, if there were any. And if his will is sufficient for that, it seems to me that it is also sufficient for the existence of my atoms.

'If anyone should say,' you continue, Sir, 'that it is God's will that a planet move circularly around its orbit, without anything aiding and conserving it in this, I say that this would be a perpetual miracle, etc'. But I could rightly mock such a philosopher, as I would mock a man who would like to pass for an architect, and who nevertheless could not make any building, even though he had all sorts of good materials suitable for this. But the best architect will do nothing without materials, just as the best philosopher will not explain the perpetual constancy of nature without atoms, which should be the materials that he must be granted. 'By the nature of things,' you say, Sir, [515/516], 'the circulating planet tends to depart from its orbit along the tangent, if nothing prevents it, and God would have to prevent this perpetually by a miracle if something did not do so naturally.' As for me, Sir, I believe that the planets could remain at a certain distance from the sun without any circulating motion, because they are supported in place by their atmospheres, as I have amply explained in the *Eclaircissements sur les conjectures physiques*, and I am of the opinion that Sir Newton, and all those before and after him of this sentiment, are mistaken when they have proposed that the planets remain in their orbits because they tend to depart along their tangent; for certainly there is no centrifugal force to consider in bodies which are in equilibrium with the matter in which they float, and which transports them around.[18] If the planets were bodies which went only by their own motion, this would be something else. 'Are you not obliged,' you say, 'to admit that [516/517] an atom *D* may

[18] This is a strange claim. The paradigm of a body having centrifugal force (introduced by Descartes) is a stone being whirled in a sling: the centrifugal endeavour of the stone is felt in the tension of the sling, which is greater the faster the stone is whirled. But according to the conceptions then current, there is an equilibrium between the centripetal and centrifugal forces involved, so that without centrifugal force there could be no equilibrium! For discussion, see Bertoloni Meli (1993, 197–9).

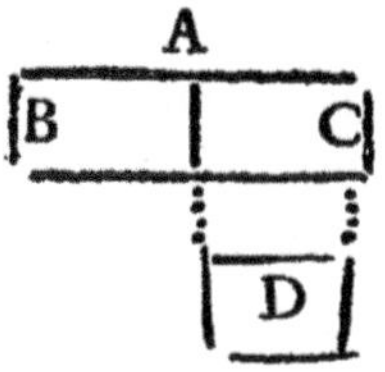

Fig. 25.1

be conceived to move against the atom *A*, in such a way that it goes directly against part *B*, etc.'[19]

Yes, without doubt, Sir, but I hold that the atom *D* could very well impel the part *C* of the atom *A* without being able to detach it from the part *B* even if it had a hundred thousand million times the speed of a launched cannonball, because it would be something contrary to the will of God, who wished the bodies we call atoms to have a perfect and insuperable hardness. Thus I hold with reason that an atom is a solid mass, and a small whole without parts, that is to say, without parts that could be detached from one another. If the body *A* were not an atom, but composed of two atoms *B* and *C*, the atom *C* could, without any difficulty, [517/518] be detached from the atom *B*, if they were not connected together by the heaviness of the atmosphere of the earth, or in some other way.

'Supposing we grant atoms,' you say, Sir, 'but are far from from admitting the void, this would not force us to resort to a first element, that is to say, a perfectly fluid matter. For why might we not replace space by a matter that has different degrees of fluidity and tenacity, as I believe is the nature of all matter?' But if we admit atoms, we would necessarily have to admit either the void or the first element, in order for it to serve the function of the void. If you want to save the motion of atoms without either the void or my first element, and form a matter with different degrees of fluidity and tenacity, you will fall into a manifest contradiction; and I do not understand, Sir, how this has been able to enter into your thought. When you call my first element a perfectly fluid matter, you are mistaken [518/519], Sir, since it is further from matter than the sky is from the Earth, and differs from it more than day differs from night. You will no doubt ask me, Sir, what then is my first element? But I will reply to you that I know nothing about it, and that it is perhaps a substance, or some thing from which the beings that we call minds are taken,

[19] Here Hartsoeker carelessly misquotes Leibniz: the atom *D* is supposed *not* to go directly against part *B*.

and which remain such as they are by the will of God; that is to say, that they continue to have life and intelligence either for a limited time, or forever. And certainly, Sir, by what demonstration could I be made to see that all that is extended must necessarily be matter, incapable of anything by itself, and that an extended being could not become mind, to have intelligence, and so forth? Since matter is incapable by itself of anything, and of any motion, and since I consider my first element as an agent and as an immaterial extension, I hold that matter has all its motion [519/520] from the first element, just as this element has all its motion from God. I have said that the universe is like an animal full of life and intelligence because I conceive that the first element could be endowed with life and intelligence under the direction of God, to whom it is a subordinate being, and to move bodies that are incapable of moving themselves. And I do not see that this would lead one to believe that God is the soul of the universe, or even the universe itself, according to the earliest philosophers. I don't know whether my first element is organic or not, nor what it must be like in order to have life or intelligence; whether for this reason it must be united to an organized body, or not, etc. But it seems to me that God would be able to grant intelligence to a portion of my first element, either for a limited time or forever, and to grant it the freedom and the power to move bodies, etc., just as we sense this same freedom and power in ourselves. I have said that there are even pieces in [520/521] the universe that develop imperfectly because I believe that there are beings subordinate to God that work continually with full freedom; but which often fall short, because their power is not infinite, and the irregularity of matter often prevents them from being successful.

'I have sufficiently satisfied,' you say, Sir, 'those who have objected to me that according to this there would no longer be *freedom*; for God, knowing what minds will freely choose in time, has accommodated bodies to this in advance.' But it appears to me that as soon as we admit that God knows what minds will choose, we must admit at the same time that they are not free, and that as soon as we maintain that they are free, and that God grants them a certain freedom for being absolute masters of their actions, he is deprived by this of his foreknowledge, and no longer knows whether they will or will not take the actions that he has left to their disposition. But I admit to you, Sir, that this matter is too far beyond my reach [521/522] to decide anything. I am with all the zeal and respect imaginable, more than anyone else in the world, Sir,

Your very humble servant, etc.
Nicolaas Hartsoeker

Dusseldorf this 13th March, 1711

Note by the editors of the *Journal de Trévoux*

We will follow up on this dispute next month. We declare, however, that we condemn what Mr Hartsoeker advances and what he insinuates: that we condemn it, I say, as much as a false system, tending toward atheism, merits being condemned. It is up to him to destroy the bad impressions that naturally arise from the views he has expressed on the eternity of matter, on his first element, on the divinity of the universe, and on the impossibility of reconciling foreknowledge and freedom. We hope that he will do so. We are not worried, moreover, that these impious conjectures hazarded in an argument might do harm; [522/523] they are weak in themselves, and Mr Leibniz easily refutes them.

26

Letter from Baron von Leibniz to Mr Hartsoeker, 12 July 1711[1]

G. W. Leibniz, *Mémoires de Trévoux*, April 1712[2]

Sir,

To respond in my turn to the honour of your letter of 13 March, I will go over its contents again. [676/677]

1. You say that nothing prevents God from having divided matter into bodies of perfect hardness, and that it was this way by itself, or by the will of God alone. But in my view, there are obstacles to this. The first is that he would have limited the subdivisions without reason. The second is that there must be a reason for the hardness, since matter is divisible, or at least, no reason prevents it from being so, and the will of God is always reasonable. I could adduce still other obstacles, but they would take me too far afield.

2. You find too little motion, Sir, in the parts of a diamond to believe that this motion would be capable to giving it this great hardness. To respond, I would say to you, Sir, that by the motions that conspire to prevent the separation of two bodies, I understand those of the fluid bodies which flow through when their movement is disturbed by the separation, and make an effort to restore it. This is why a small quantity of gunpowder has so much force, and even a force that exceeds what is needed to break a diamond of the same weight as the powder, for the motion of the surrounding bodies must be joined to that of the parts of the powder, otherwise one would have difficulty accounting for this great explosion.

[1] From the French: '*Lettre de Monsieur Le Baron de Leibnits à M. Hartsoeker, le 12. de Juillet 1711*', *Mémoires de Trévoux* (more formally, *Mémoires pour l'Histoire des Sciences & des Beaux-Arts*), April 1712, 676–9; republished in the *Journal des Sçavans*, January 1713, 73–6. (GP III 527–28). Translated by Tzuchien Tho.

[2] This was Leibniz's reply to Hartsoeker's response of 13 March 1711, dated by him 'Hanover, this 9th of July, 1711.' In it he reiterates his points about the unintelligibility of perfect hardness and action at a distance, defends his explanation of cohesion in terms of conspiring motions, and critiques Hartsoeker's understanding of organic matter.

Leibniz: Journal Articles on Natural Philosophy. Richard T. W. Arthur, Oxford University Press.
DOI: 10.1093/oso/9780192843531.003.0027

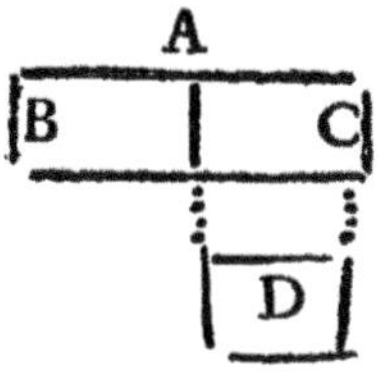

Fig. 25.1

3. I said that if God wanted a planet to circulate in its orbit [677/678] without anything helping it in this, or if he prevented it from going off along the tangent, this would be a perpetual miracle. I say this again, and you do not respond to me on this issue, Sir, when you say that the equilibrium of matter in which the planets are floating prevents them from moving away. For in saying that you are supposing something which prevents them from doing this, against the supposition in question. So what I said stands, and the primitive hardness of a body would consist in a miracle, similar to that which conserves the planets in their orbit without employing any further contributing factor.

4. You take recourse to the will of God to account for why *D* cannot carry off *C* without carrying away *B*; but since you do not recognize anything that could serve to explain how this will is carried out, you leave the natural and save yourself in a miracle, exactly like someone who explains the motion of the planets in their orbit only through the will of God.

5. You do not say, Sir, why we [678/679] should admit only two kinds of matter, one perfectly hard, that is to say, that of the atoms, and another perfectly fluid; and why it is not possible for there to be intermediate kinds of matter, whose hardness or fluidity could be surpassed?

6. You do not, Sir, wish your first element or your perfect fluid to be called matter. Is this not a dispute about a word? It is an extended and resistant body. It is customary to call that matter, and your fluid must be resistant, because it is capable of moving atoms.

7. But you say that it is acting. So: it is because God first impressed force on it, and if God had likewise impressed force or motion on the atoms, just as he has impressed it on your fluid, would they be less material for that? There is even reason to believe that God has impressed force on all bodies.

8. If God should come to put a soul in a portion of matter, or in an extended thing, he will endow it with organs, otherwise it will not act in an orderly way.

9. It has been sufficiently shown elsewhere how freedom is in no way opposed to foreknowledge, or to certainty, and I refer you to what has already been said on this issue. I am, etc.

Bibliography

1. Primary Sources*

Ango, Pierre. 1682. *L'optique, divisée en trois livres*. Paris: Estienne Michallet.

Archimedes. 1897. 'On the Equilibrium of Planes', in Heath, T.L. *The Works of Archimedes* (1897). London: Cambridge University Press.

Barrow, Isaac. 1670. *Lectiones Geometricæ: in quibus (praesertim) generalis curvarum linearum symptomata declarantur*. London: Godbid.

Barrow, Isaac. 1916. *The Geometrical Lectures of Isaac Barrow*. Translation and notes by James M. Child. London: Open Court.

Bernoulli, Jacob. 1685. '*Problème proposé par M. Bernoulli*', *Journal des sçavans*, 26 August 1685, 314.

Bernoulli, Jacob. 1686. '*Dubium Circa Causam Gravitatis a rotatione Vorticis Terreni petitam*', *Acta eruditorum*, May, 91–5.

Bernoulli, Jacob. 1690a. '*Analysis problematis antehac propositi, de inventione lineae descensus a corpore gravi percurrendæ uniformiter, sic ut temporibus æqualibus æquales altitudines emetiatur: & alterius cujusdam Problematis Propositio*', *Acta eruditorum*, May, 217–19.

Bernoulli, Jacob. 1690b. '*Quaestiones nonnullæ de usuris, cum solutione problematis de sorte alearum*', *Acta eruditorum*, May, 219–23.

Bernoulli, Jacob. 1691a. '*Specimen calculi differentialis in dimensione Parabolæ helicoïdis ubi de flexuris curvarum in genere earundem evolutionibus aliisque*', *Acta eruditorum*, January, 13–23.

Bernoulli, Jacob. 1691b. '*Specimen alterum calculi differentialis in dimetienda Spirali Logarithmica, Loxodromiis Nautarum, & Areis Triangulorum Sphæricorum: una cum Addiamento quodam ad Problema Funicularium, aliisque*', *Acta eruditorum*, June, 282–90.

Bernoulli, Jacob. 1691c. '*Demonstratio Centri Oscillationis ex Natura Vectis, reperta occasione eorum, quæ super hac materia in Historia Literaria Roterdamensi recensentur, articulo 2. mens. Jun. 1690*', *Acta eruditorum*, June, 317–21.

Bernoulli, Jacob. 1693. '*Curvæ Diacausticæ, eorum Relatio ad Evolutas, aliasque nova his affinia. Regula pro resistentiis, quas Figuræ in Fluido, mote patiuntur &c. per I. B.*', *Acta eruditorum*, May, pp. 244–56.

Bernoulli, Jacob. 1694a. '*Curvatura laminæ elasticæ. Ejus Identitas cum Curvatura Linei a pondere inclusi fluidi expansi. Radii circulorum osculantium in terminis simplicissimis exhibiti; una cum nobis quibusdam Theorematis huc pertinentibus &c.*', *Acta eruditorum*, June 1694 (wrongly listed as June 1692 on the first two pages), 262–76.

* We have not included here the articles contained in this volume, whose details are given at the beginning of each piece.

Bernoulli, Jacob. 1694b. '*Solutio Problematis Leibnitiani Curva Accessus & Recessus æquabilis a puncto dato, mediante rectificatione Curvæ Elasticæ*', *Acta eruditorum*, June 1694, 276–80.

Bernoulli, Jacob. 1695. '*Explicationes, annotationes et additiones*', *Acta eruditorum*, December, pp. 537–53.

Bernoulli, Johann. 1691. '*Solutiones Problematis Funicularii, exhibita a Johanne Bernoulli, Basil. Med. Cand.*', *Acta eruditorum*, June, 274–6.

Blondel, François. 1683. *L'Art de jetter les bombes*. Paris: Nicolas Langlois.

Borelli, Giovanni Alfonso. 1667. *De vi percussionis*. Bologna: Jacobus Montius.

Borelli, Giovanni Alfonso. 1670. *De motionibus naturalibus a gravitate pendentibus*. Bologna: Dominici Ferri.

Boyle, Robert. 1674. 'Of the Excellency and Grounds of the Corpuscular or Mechanical Philosophy', in 'The Excellency of Theology, Compar'd with Natural Philosophy' *Philosophical Transactions of the Royal Society of London*.

Bruno, Giordano. 1584. *De l'infinito universo e mondi*. (Stampato in Venezia). London.

Craig, John. 1685. *Methodus figurarum lineis rectis et curvis comprehensarum quadraturas determinandi*. London: Moses Pitt.

Dechales (or De Chales), Claude François Milliet. 1674. *Cursus seu mundus mathematicus*. Lyon: Ex Officina Anissoniana.

Descartes, René. 1964–76. *Oeuvres de Descartes*, 12 vols, Nouvelle présentation, ed. Charles Adam and Paul Tannery. Paris: J. Vrin. (abbrev. AT)

Fabri, Honoré. 1646. *Tractatus Physicus De Motu Locali*. Lyon: Johannes Champion.

Fatio de Duillier, Nicolas. 1689. 'Réponse de M. N. Fatio de Duillier, de la Societé Roiale de Londre, à un Écrit de M. de T. qui a été publié dans le Tome X. de la Bibliotheque Universelle: touchant une maniére de déterminer les tangentes des lignes courbes, qui se peuvent décrire par des fils', *Bibliothèque Universelle et Historique*, XIII, April, 46–76.

Fatio de Duillier, Nicolas. 1699. *Lineæ brevissime descensus investigatio geometrica duplex, cui addita est investigatio gemotrica solidi rotundi in quod minima fiat resistentia*. London: J. Taylor.

Galileo Galilei. 1638. *Discorsi e dimostrazioni matematiche, intorno a due nuove scienze attenenti alla mecanica & i movimenti locali*. Leiden: Elsevier.

Galileo Galilei. 1974. *Discourses and mathematical demonstrations concerning two new sciences pertaining to mechanics and local motions*. Trans. Stillman Drake. Madison: University of Wisconsin Press.

Gregory, David. 1702. *Astronomiæ physicæ & geometricæ elementa*. Oxford: Sheldonian Theatre.

Gregory, James. 1668. *Vera circuli et hyperbolae quadratura*. Padua: Paoli Frambotti.

Hartsoeker, Nicolaas. 1706. *Conjectures physiques*. Amsterdam: Henri Desbordes.

Hartsoeker, Nicolaas. 1710. *Eclaircissements sur les Conjectures physiques*. Amsterdam: Pierre Humbert.

Huygens, Christiaan. 1669. 'A summary Account of the Laws of Motion', *Philosophical Transactions*, 46, 925–8.

Huygens, Christiaan. 1673. *Horologium Oscillatorium sive de motu pendulorum ad horologia adapto demonstrationes geometricae*. Paris: F. Muguet.

Huygens, Christiaan. 1687. '*Solution du problème proposé par M. L[eibniz] dans les Nouvelles de la Republique des Lettres, du mois de Septembre 1687*', *Nouvelles de la Republique des Lettres*, October, pp. 1110–11.

Huygens, Christiaan. 1690. *Traité de la Lumière* (including *Discours de la cause de la pesanteur* as supplement). Leiden: Pierre van der Aa.

Huygens, Christiaan. 1691. '*Christiani Hugenii, Dynaste in Zülichem, solutio ejusdem Problematis*', *Acta eruditorum*, June, 281–2.

Huygens, Christiaan. 1698. *Cosmotheoros*. The Hague: Adriaan Moetjens.

Huygens, Christiaan. 1888–1950. *Œuvres complètes*. The Hague: Martinus Nijhoff (abbrev. OCH).

Huygens, Christiaan. 1986. *The Pendulum Clock or Geometrical Demonstrations Concerning the Motion of Pendula as Applied to Clocks*. Trans. Richard J. Blackwell. Ames: Iowa State University Press.

Jaquelot, Isaac. 1705. *Conformité de la foi avec la raison: ou défense de la religion, contre les principales difficultez répandues dans le Dictionaire historique et critique de Mr. Bayle*. Amsterdam: Henry Desbordes and Daniel Pain.

Kepler, Johannes. 1604. *Paralipomena ad Vitellione*. Frankfurt: Claude Marne and the heirs of Jean Aubry.

Kepler, Johannes. 1609. *Astronomia Nova*. Heidelberg: G. Voegelinus.

Kepler, Johannes. 1937. *Gesammelte Werke*. Ed. Walther von Dyck and Max Caspar. Munich: C. H. Beck.

Lamy, François. 1697. *Connoissance de soi-même*. Paris: André Pralard.

Leibniz, G. W. 1675. '*Extrait d'une Lettre de Mr. Leibniz à l'Auteur du Journal, touchant le principe de justesse des Horloges portatives de son Invention*', *Journal des Sçavans*, March 1675, 93–6.

Leibniz, Gottfried W. 1677a. '*Extrait d'une lettre de M. Leibniz à l'Auteur du Journal, ecrit à Hanovre le 18. Juin 1677, contenant la relation et la figure d'un chevreuil coiffé d'une manière fort extraordinaire*', *Journal des Sçavans*, 5 July 1677, 165–7.

Leibniz, Gottfried W. 1677b. '*Le phosphore de M. Krafft ou liqueur et terre sèche de sa composition qui jettent continuellement de grands éclats de lumière*', *Journal des Sçavans*, 2 August 1677, 190–1.

Leibniz, G. W. 1681. '*Extrait de deux lettres écrites à l'auteur du Journal, l'une d'Hanovre par M. de Leibnitz, Conseiller de S.A.M. le Duc d'Hanover, touchant une expérience considérable d'une eau fumante, et l'autre d'Oxford, par M. Hansen*', *Journal des Sçavans*, February, 46.

Leibniz, G. W. 1682. '*De vera proportione circuli ad quadratum circumscriptum in numeris rationalibus expressa*', *Acta eruditorum*, February, 41–6.

Leibniz, G. W. 1684a. '*De dimensionibus figurarum inveniendis*', *Acta eruditorum*, May, 233–6.

Leibniz, G. W. 1684b. '*Nova methodus pro maximis et minimis, itemque tangentibus, quae nec fractas, nec irrationales quantitates moratur, et singulare pro illis calculi genus*', *Acta eruditorum*, October, 467–73.

Leibniz, G. W. 1684c. '*Meditationes de Cognitione, Veritate & Ideis*', *Acta eruditorum*, November, 537–42.

Leibniz, G. W. 1684d. '*Additio ad Schedam De dimensionibus figurarum inveniendis*', *Acta eruditorum*, December, 585–7.

Leibniz, G. W. 1685. '*Demonstratio Geometrica Regulae apud Staticos receptae de momentis gravium in planis inclinatis, nuper in dubium vocatae, et solutio casus elegantis, in Actis Novembr. 1684 pag. 512 propositi, de globo duobus planis angulum rectum facientibus simul incumbente, quantum unumquodque planorum prematur, determinans*', *Acta eruditorum*, November, 501–5.

Leibniz, G. W. 1686a. '*Meditatio nova de natura anguli contactus et osculi, horumque usu in practica mathesi, ad figuras faciliores succedaneas difficilioribus substituendas*', *Acta eruditorum*, June, 289–92.

Leibniz, G. W. 1686b. '*De geometria recondita et analysi indivisibilium atque infinitorum, addenda his qua dicta sunt in Actis a. 1684, Maji, p. 233; Octob. p. 264; Decemb. p. 586*', *Acta eruditorum*, June, 292–300.

Leibniz, G. W. 1687. '*Extrait d'une Lettre de M. L. sur un Principe Générale, utile à l'explication des loix de la nature, par la considération de la Sagesse Divine; pour servire de réplique à la réponse du R. P. M, a.a.O.*', *Nouvelles de la République des Lettres*, July, 744–53.

Leibniz, G. W. 1691. '*Quadratura arithmetica communis sectionum conicarum quæ centrum habent, indeque ducta trigonometria canonica ad quantamcumque in numeris exactitudinem a tabularum necessitate liberata: cum usu speciali ad lineam rhomborum nauticam, aptatumque illi planisphærium*', *Acta eruditorum*, April, 178–82.

Leibniz, G. W. 1694. '*Constructio propria Problematis de Curva Isochrona Paracentrica. Ubi et generaliora quædam de natura et calculo differentiali osculorum, et de constructione linearumn transcendentium, una maxime geometrica, altera mechanica quidem, sed generalissima. Accessit modus reddendi inventiones transcendentium linearum universales, ut quemvis casum comprehendant, et transeant per punctum datum*', *Acta eruditorum*, August 1694, 364–75.

Leibniz, G. W. 1695. '*Responsio ad nonnullas difficultates a Dn. Bernardo Niewnetijt circa Methodum differentialem seu Infinitesimalem motas*', *Acta eruditorum*, July, 310–16.

Leibniz, Gottfried W. 1700. *Responsio ad Dn. Nic. Fatii Duillerii Imputationes. Accessit nova artis analyticæ promotio specimine indicata; dum designatione per numeros assumtitios loco litterarum, algebra ex combinatoria arte lucem capit*, *Acta eruditorum*, May, 198–208.

Leibniz, G. W. 1749. *Protagea: sive, De prima facie telluris et antiquissimæ historiæ vestigiis in ipsis naturæ monumentis dissertatio*. Ed. Christian Ludwig Scheidt, Göttingen.

Leibniz, G. W. 1768. *G. G. Leibnitii Opera Omnia*. 6 vols. Ed. Louis Dutens. Geneva: Fratres de Tournes.

Leibniz, G. W. 1849–63. *Leibnizens Mathematische Schriften*. Ed. C. I. Gerhardt. Berlin and Halle: Asher and Schmidt, 1849–63. 7 vols. (abbrev. GM).

Leibniz, G. W. 1875–90. *Die Philosophischen Schriften von Gottfried Wilhelm Leibniz*. Ed. C. I. Gerhardt. Berlin: Weidmann. 7 vols. (abbrev. GP).

Leibniz, G. W. 1923–. *Sämtliche Schriften und Briefe*. Ed. Akademie der Wissenschaften. Berlin: Weidmann. 7 vols. (abbrev. A).

Leibniz, G. W. 1956. *The Leibniz-Clarke Correspondence*. Ed. H. G. Alexander. Manchester: Manchester University Press.

Leibniz, G. W. 1976. *Philosophical Papers and Letters*. Second Edition. Transl. and Ed. with an Introduction by Leroy E. Loemker. Dordrecht/Boston: D. Reidel.

Leibniz, G. W. 1994. *La réforme de la dynamique. De Corporum concursu (1678) et autres textes inédits*. Ed. and Transl. with commentary by Michel Fichant. Paris: Vrin.

Leibniz, G. W. 1995. *La naissance du calcul différentiel: 26 articles des* Acta eruditorum, ed. and transl. Marc Parmentier. Paris: Vrin.

Leibniz, G. W. 2001. *The Labyrinth of the Continuum: Writings of 1672 to 1686*. Selected, edited, and translated, with an introduction by R. T. W. Arthur. New Haven: Yale University Press.

Leibniz, G. W. 2005. *Essais scientifiques et philosophiques. Les articles publiés dans les journaux savants*. Recueillis par Antonio Lamarra et Roberto Palaia. Hildesheim/Zürich/New York: Georg Olms. (abbrev. ESP).

Leibniz, Gottfried W. 2007. *The Leibniz-Des Bosses Correspondence*. Transl., Ed., and with an Introduction by Brandon C. Look and Donald Rutherford. New Haven & London: Yale University Press.

Leibniz, G. W. 2011. *Die mathematischen Zeitschriftenartikel*. Ed. and Transl. Heinz-Jürgen Hess and Malte-Ludolf Babin. Hildesheim/New York: Olms.

Leibniz, Gottfried W. 2013. *The Leibniz-De Volder Correspondence*. With selections from the Correspondence Between Leibniz and Johann Bernoulli. Transl., Ed., and with an Introduction by Paul Lodge. New Haven & London: Yale University Press.

Leibniz, Gottfried W. 2015. *G. W. Leibniz, Interrelations between Mathematics and Philosophy*, eds. Norma B. Goethe, Philip Beeley, and David Rabouin. Dordrecht: Springer.

L'Hôpital, Guillaume, Marquis de. 1693a. *Mémoires de l'Académie Royale des Sciences*, 30 June 1693.

L'Hôpital, Guillaume, Marquis de. 1693b. *Problematis, a Joh. Bernoullio in hisce Actis Mense Majo pag. 235 propositi, Solutio, a Dn. Marchione Hospital in literis ad Dn. Bernoullium d. 27 Junii Exhibita, Acta eruditorum*, September 1693, 398–9.

L'Hôpital, Guillaume, Marquis de. 1696. *Analyse des Infiniment Petits pour l'Intelligence des Lignes Courbes*. Paris: Montalant.

Locke, John. 1690. *An Essay Concerning Humane Understanding*. Second Edition. London: printed by Eliz. Holt, for Thomas Basset. Ed. with an introduction by P. H. Nidditch. Oxford: Clarendon Press, 1975.

Malebranche, Nicolas. 1674–75. *De la recherche de la verité*. Paris: Michel David.

Malebranche, Nicolas. 1692. *Lois de la communication des mouvements*. In vol. 2 of *Oeuvres complètes de Malebranche*, 20 vols. Ed. André Robinet. Paris: J. Vrin, 1958–1984.

Mariotte, Edme. 1676. *De la nature de l'air*. Republished in *Essais de Physique, ou mémoires pour servir à la science des choses naturelles*. Paris: E. Michallet, 1679.

Mariotte, Edme. 1684. *Traité du mouvement des eaux*. In *Traité du mouvement des eaux et des autres corps fluides, divisé en V parties, par feu M. Mariotte, mis en lumière par les soins de M. de La Hire* (Paris: E. Michallet, 1686).

Mariotte, Edme. 1703. Traité de la percussion ou choc des corps, dans lequel les principales règles du mouvement, contraires à celles que Mr. Descartes et quelques autres modernes ont voulu établir, sont démontrées par leurs véritables causes. Paris: Étienne Michallet.

Monconys, Balthasar de. 1665. *Journal des Voyages de Monsieur de Monconys*. Lyon: Horace Boissat and George Remeus.

Newton, Isaac. 1687. *Philosophiae Naturalis Principia Mathematica*. London: Joseph Streater.

Newton, Isaac. 1704. *Opticks: or, a treatise of the reflexions, refractions, inflexions and colours of light*. London: Samuel Smith and Benjamin Walford.

Newton, Isaac. 1999. *The* Principia*: Mathematical Principles of Mathematical Philosophy*. Ed. and Transl. I. Bernard Cohen and Anne Whitman. Berkeley: University of California Press.

Nieuwentijt, Bernard. 1694. *Considerationes circa analyseos ad quantitates infinite parvas applicatae principia, et calculi differentialis usum in resolvendis problematibus geometricis*. Amsterdam: Johannes Wolters.

Nieuwentijt, Bernard. 1695. *Analysis infinitorum, seu curvilineorum proprietates ex polygonorum natura deductae*. Amsterdam: Johannes Wolters.

Pardies, Ignace Gaston. 1670. *Discours du mouvement locale, avec les remarques sur le movement de la lumière*. Paris: Edme Martin.

Roberval, Gilles Personne de. 1644. *Aristarchi Samii de Mundi Systemate*. Paris: Antonium Bertier.

Saint-Vincent, Grégoire de. 1647. *Opus geometricum quadraturae circuli et sectionum coni decem libris comprehensum*. Antwerp: Jan van Meurs & Jacob van Meurs.

Spleiss, Stephan. 1679. *De visionis distinctissimae loco. Observatio CXCI* in *Miscellanea curiosa medico-physica academiæ naturæ curiosorum*. Jena: Samuel Krebs.

Thévenot, Melchisédech. 1681. *Recueil de voyages de Mr Thévenot*. Paris: Estienne Michallet.

Tschirnhaus, Ehrenfried Walther von. 1687. *Medicina Mentis, sive Artis Inveniendi Praecepta Generalia*. Second edition 1695: Leipzig, J. Thomas Fritsch.

Tschirnhaus, Ehrenfried Walther von. 1690a. '*Methodus Curvas Determinandi, quæ formantur a radiis reflexis, quorum incidentes ut paralleli considerantur*', *Acta eruditorum*, February 1690, 68–73.

Tschirnhaus, Ehrenfried Walther von. 1690b. '*Curva geometrica, quae seipsam sui evolutione describit, aliasque insignes proprietates obtinet, inventa a D. T.*', *Acta eruditorum*, April 1690, 169–72.

Varignon, Pierre. 1700. 'Des Forces Centrales ou des pesanteurs nécessaires aux Planetes pour leur faire décrire les orbes qu'on leur a supposés jusqu'ici', *Mémoire de l'Académie Royale des Sciences*, 224–36.

Varignon, Pierre. 1701. 'Autre Regle Generale des Forces Centrales', *Mémoire de l'Académie Royale des Sciences*, 20–39.

Vossius, Isaac. 1662. *De lucis naturae et proprietate*. Amsterdam: Ludovic and Daniel Elzevir.

2. Secondary Sources

Aiton, Eric J. 1960. 'The Celestial Mechanics of Leibniz', *Annals of Science*, **16**, 2, 65–82.

Aiton, Eric J. 1972a. 'Leibniz on motion in a resisting medium', *Arch. Rational Mech.* **9**, 257–74.

Aiton, E. J., 1972b. *The Vortex Theory of Planetary Motions*. London: MacDonald.

Alexander, H. G. 1956. *The Leibniz-Clarke Correspondence*. Ed. with introduction and notes by H. G. Alexander. Manchester and New York: Manchester University Press.

Antognazza, Maria Rosa. 2009. *Leibniz: An Intellectual Biography*. Cambridge: Cambridge University Press.

Antognazza, Maria Rosa. 2018. *The Oxford Handbook of Leibniz*. Oxford: Oxford University Press.

Arthur, Richard T. W. 1995. "Newton's Fluxions and Equably Flowing Time," *Studies in History and Philosophy of Science*, **28**, no. 2 (June), 323–51.

Arthur, Richard T. W. 2016. 'On the Mathematization of Free Fall: Galileo, Descartes and a History of Misconstrual', 81–111 in *The Language of Nature*, volume 20 of *Minnesota Studies in Philosophy of Science*, ed. Geoffrey Gorham, Benjamin Hill, Edward Slowik and C. Kenneth Waters (Minneapolis: University of Minnesota Press, 2016).

Arthur, Richard T. W. 2018. *Monads, Composition and Force: Ariadnean Threads through Leibniz's Labyrinth*. Oxford: Oxford University Press.

Arthur, Richard T. W. 2021. *Leibniz on Time, Space, and Relativity*. Oxford: Oxford University Press.

Auger, Léon. 1957. 'Les idées de Roberval sur le système du monde', *Revue d'Histoire des sciences*, **10**, 3, 226–34.

Beeley, Philip. 2018. 'Early Physics', pp. 290–303 in Antognazza (2018).

Beeley, Philip. 2020. '"There are great alterations in the geometry of late". The rise of Isaac Newton's early Scottish circle', *British Journal for the History of Mathematics*, **35**, 1, 3–24.

Bertoloni Meli, Domenico. 1988. 'Leibniz on the Censorship of the Copernican System', *Studia Leibnitiana*, **20**, 1, 19–42.

Bertoloni Meli, Domenico. 1993. *Equivalence and Priority: Newton versus Leibniz; including Leibniz's unpublished manuscripts on the* Principia. Oxford: Oxford University Press.

Bertoloni Meli, Domenico. 2006. *Thinking with Objects: The Transformation of Mechanics in the Seventeenth Century*. Baltimore: Johns Hopkins University Press.

Bos, H. J. M. 1974–75. Differentials, Higher-Order Differentials and the Derivative in the Leibnizian Calculus, *Archive for History of Exact sciences*, **14**, 1–90.

Coopersmith, Jennifer. 2017. *The Lazy Universe*. Oxford: Oxford University Press.

Costabel, Pierre. 1960. *Leibniz et la dynamique Les textes de 1692*. Paris, Hermann.

Damerow, Peter, Gideon Freudenthal, Peter McLaughlin, and Jürgen Renn. 1992. *Exploring the Limits of Preclassical Mechanics*. New York: Springer.

Des Chene, Dennis. 1996. *Physiologia: Natural Philosophy in Late Aristotelian and Cartesian Thought*. Ithaca and London: Cornell University Press.

Duchesneau, François. 1994. *La dynamique de Leibniz*. Paris: Vrin.

Eagles, Christina. 1977. *The Mathematical Work of David Gregory, 1659–1708*. Ph.D. Thesis. Edinburgh University.

Feingold, Mordechai (ed.). 1990. *Before Newton*. Cambridge: Cambridge University Press.

Feingold, Mordechai. 1993. 'Newton, Leibniz, and Barrow too, an attempt at a reinterpretation', *Isis*, **84**, 310–38.

Garber, Daniel. 2004. 'Leibniz and Fardella: Body, Substance and Idealism', pp. 123–40 in Lodge (ed.) (2014).

Garber, Daniel and Tzuchien Tho. 2018. 'Force and Dynamics', pp. 304–30 in Antognazza (ed.) (2018).

Guicciardini, Niccolò. 1999. *Reading the* Principia: *The Debate on Newton's Mathematical Methods for Natural Philosophy from 1687 to 1736*. Cambridge: Cambridge University Press.

Guicciardini, Niccolò. 2009. *Isaac Newton on Mathematical Certainty and Method*. Cambridge, MA: MIT Press.

Jullien, Vincent and André Charrak. 2002. *Ce que dit Descartes touchant la chute des graves*. Paris: Presses Universitaires de Septentrion.

Lodge, Paul. 2004. *Leibniz and his Correspondents*. Cambridge: Cambridge University Press, 2004.

McDonough, Jeffrey K. 2018. 'Optics', 425–37 in Antognazza (2018).

McDonough, Jeffrey K. 2022. *A Miracle Creed. The Principle of Optimality in Leibniz's Physics and Philosophy*. Oxford: Oxford University Press.

Mahoney, Michael. 1990. 'Barrow's Mathematics: Between ancients and moderns', pp. 179–249 in Feingold (ed.) (1990).

O'Hara, James. (forthcoming). *Leibniz's Correspondence in Science, Technology and Medicine (1676–1701): Core Themes and Core Texts*. Leiden: Brill.

Probst, Siegmund. 2015. 'Leibniz as Reader and Second Inventor: The Cases of Barrow and Mengoli', pp. 111–34 in Leibniz (2015).

Rabouin, David. 2013. '"*Analytica Generalissima Humanorum Cognitionum*". Some Reflections on the Relationship between Logical and Mathematical Analysis in Leibniz', *Studia Leibnitiana*, **45**, 1, 109–30.

Robinet André. 1958. 'L'abbé de Catelan, ou l'erreur au service de la verité'. *Revue d'histoire des sciences et de leurs applications*, **11**, 4, 289–301.

Roero, Clara Silvia. 1990. 'Leibniz and the Temple of Viviani: Leibniz's prompt reply to the challenge and the repercussions in the field of mathematics', *Annals of Science*, **47**, 5, 423–43.

Sabra, A. I. 1981. *Theories of Light from Descartes to Newton*. Cambridge: Cambridge University Press.

Schepers, Heinrich. 2018. 'Space and Time', 410–22 in Antognazza (2018).

Woolhouse, R. S. 2001. 'Leibniz and François Lamy's *De la Connaissance de soi-même*', *Leibniz Review*, 11, 65–70.

Woolhouse, Roger S. and Richard Francks. 1997. *Leibniz's 'New System' and Associated Contemporary Texts*. Oxford: Oxford University Press.

Index of Names

Subject Index